Lecture Notes in Mathematics 1813

Editors:
J.-M. Morel, Cachan
F. Takens, Groningen
B. Teissier, Paris

Subseries:
Fondazione C.I.M.E., Firenze
Adviser: Pietro Zecca

Springer
Berlin
Heidelberg
New York
Hong Kong
London
Milan
Paris
Tokyo

L. Ambrosio   L. A. Caffarelli   Y. Brenier
G. Buttazzo   C. Villani

Optimal Transportation and Applications

Lectures given at the
C.I.M.E. Summer School
held in Martina Franca, Italy,
September 2–8, 2001

Editors: L. A. Caffarelli
         S. Salsa

Fondazione
C.I.M.E.

Authors and Editors

Luigi Ambrosio
Scuola Normale Superiore
Piazza dei Cavalieri 7
56126 Pisa, Italy
e-mail: luigi@ambrosio.sns.it

Luis A. Caffarelli
Department of Mathematics
University of Texas at Austin
Austin, TX78712-1082, USA
e-mail: caffarel@math.utexas.edu

Yann Brenier
Laboratoire J. A. Dieudonné
Université de Nice-Sophia Antipolis
Parc Valrose
06108 Nice, France
e-mail: brenier@math.unice.fr

Giuseppe Buttazzo
Department of Mathematics
University of Pisa
Via Buonarroti 2
56127 Pisa, Italy
e-mail: buttazzo@dm.unipi.it

Cedric Villani
UMPA
École Normale Supérieure de Lyon
69364 Lyon Cedex 07, France
e-mail: cvillani@umpa.ens-lyon.fr

Sandro Salsa
Department of Mathematics 'F. Brioschi'
Politecnico di Milano
Piazza Leonardo da Vinci 32
20133 Milano, Italy
e-mail: sansal@mate.polimi.it

Cataloging-in-Publication Data applied for
Bibliographic information published by Die Deutsche Bibliothek

Die Deutsche Bibliothek lists this publication in the Deutsche Nationalbibliografie;
detailed bibliographic data is available in the Internet at http://dnb.ddb.de

Mathematics Subject Classification (2000): 35J20, 35J70, 35K65, 35L70, 35Q60, 35Q75, 49K20, 49K99, 60F99, 52A99, 74P99, 82C70, 82C99

ISSN 0075-8434
ISBN 3-540-40192-X Springer-Verlag Berlin Heidelberg New York

Springer-Verlag Berlin Heidelberg New York a member of BertelsmannSpringer
Science + Business Media GmbH

http://www.springer.de

© Springer-Verlag Berlin Heidelberg 2003
Printed in Germany

Typesetting: Camera-ready TeX output by the authors

SPIN: 10931677   41/3142/du - 543210 - Printed on acid-free paper

Preface

Optimal transportation is an old subject.

In fact the issue was raised by Monge (the Monge-Ampère equation), then rediscovered by Kantorovich in the context of economics and it is an important topic within probability (the Wasserstein metric).

It concerns, roughly, how to transport a mass (probability density) from one location (and distribution) to another, in such a way as to keep the transportation cost to a minimum - a very natural and reasonable problem. But, it was in the late '80s that the basic mathematical problem associated to it, and its connection with Monge-Ampère type equations, started to surface in very different environments: in metereology, in the theory of front formation, in the discretization of Euler equations by some sort of Lagrangian approach, in the theory of lubrication, the study of rates of decay for nonlinear evolution equations, etc..

It was found to be a lagrangian version of the div-curl decomposition, to have connection with sand pile dynamics, with statistical mechanics and many other fields.

We feel, thus, that this is a very appropriate moment to bring together a group of researchers in the field, with different views and perspectives in the topic, to create a basic "guide" to those young researchers, that may find this theory challenging and useful.

The C.I.M.E. course on Optimal Transportation and Applications, held in Martina Franca (Italy) from September 2 to September 8, 2001, was designed precisely to fulfil this purpose.

The school was organized into the following five courses:

G. Buttazzo: *Shape optimization problems through the Monge-Kantorovich equation*

Y. Brenier: *Geometric PDEs related to fluids and plasmas*

L. A. Caffarelli: *The Monge-Ampère Equation, Optimal Transportation and periodic media*

L. C. Evans: *Optimal Transportation*

C. Villani: *Mass transportation tools for dissipative PDEs*

We are pleased to express our appreciation to the speakers for their beautiful lectures. The present volume records and completes the material presented in the courses listed above, with an important and original additional contribution from Ambrosio and Pratelli:

L. Ambrosio and A. Pratelli: *Existence and Stability Results in the L^1 Theory of Optimal Transportation.*

Many researchers from Italy and abroad attended the courses; we thank them for their active contribution to the success of the school. We would also like to thank the C.I.M.E. Scientific Committee for the invitation to organize the School in Martina Franca.

Luis A. Caffarelli
Sandro Salsa

CIME's activity is supported by:

Ministero dell'Università Ricerca Scientifica e Tecnologica, COFIN '99;
Ministero degli Affari Esteri - Direzione Generale per la Promozione e la Cooperazione - Ufficio V;
Consiglio Nazionale delle Ricerche;
E.U. under the Training and Mobility of Researchers Programme;
UNESCO-ROSTE, Venice Office.

Contents

The Monge-Ampère Equation and Optimal Transportation, an elementary review

Luis Caffarelli

Department of Math, UT Austin, Austin, TX 78712
caffarel@fireant.ma.utexas.edu

1 Optimal Transportation

Optimal transportation can be better described in the discrete case:

We are given "goods" sitting at k different locations, x_j, in $\mathbb{R}^n$, and we want to transport them to k new locations y_j.

We do not care which goods go to which point, and transporting them from x_i to y_j incurs a cost $C(x_i - y_j)$ (think of $C_p(x - y) = \frac{1}{p}|x - y|^p$).

We want to choose a delivery scheme $y(x)$ that would minimize the total cost:

$$J(y) = \sum_j C\big(y(x_j) - x_j\big) ,$$

among all admissible transportation plans $y(x)$.

Of course, everything being finite, such a problem has a solution $y_0(x)$,

$$J(y_0(x)) \leq \min \sum C(y_0(x_j) - x_j) \tag{1}$$

Clearly, this imposes some geometric condition on the map. For instance, suppose that $C = C_p$ (and in particular rotationally invariant).

If we take two points x_1, x_2 and their images $y_0(x_1), y(x_2)$ we may wonder what does it mean to switch them (that would increase cost). We can, for instance, take a system of coordinates where $x_1 = 0$, $x_2 = \lambda e_1$. Then, $y_0(x_1)$, $y_0(x_2)$ can be rotated with respect to this axis to make the configuration coplanar without changing cost.

This reduces the question to a problem in the plane and we see that for each position $y_0(x_1) = \alpha e_1 + \beta e_2$, $y_0(x_2)$ is forced to stay in some predetermined region above $y_0 = Ry_0$. That is, the map has to have some monotonicity.

For instance, in the case $p = 1$ (the usual Euclidean distance) we see that the vectors from x_i to y_{0_i} should not cross. For $p = 2$, instead the map $y_0(x)$ has to be monotone, i.e.,

$$\langle y(x_1) - y(x_2), x_1 - x_2 \rangle \geq 0 .$$

2 The continuous case:

In the continuous case, instead of having two finite families of locations, we are given two "goods densities", $f(x)$ and $g(y)$; that is we have our goods "spreaded" through $\mathbb{R}^n$ with density $f(x)$ and we want to reorganize them, so they become "spreaded" with density $g(y)$. (There is an obvious compatibility condition $\int f = \int g$.)

So, heuristically, we are looking for a map, $y_0(x)$, that will reorganize f into g, and that, among all "admissible" maps will minimize the transportation cost

$$J(y) = \int C(y - x) f(x) \, dx$$

Admissible means that for any set A, the total f-mass of A, be identical to the g-mass of its image, or infinitesimally (if such a thing were allowed)

$$g(y(x)) \det D_x y = f(x) \ .$$

A weak way of expressing admissibility is that, for any continuous test function $\eta(y)$

$$\int \eta(y) g(y) \, dy \ \text{``="} \int \eta(y(x)) g(y(x)) \det y_x(x) \, dx$$

be actually equal to

$$\int \eta(y(x)) f(x) \, dx \ .$$

That is, we "allow" ourself to replace the formal term

$$g(y(x)) \det D_x y \ \text{by} \ f(x) \ .$$

A still "weaker" formulation of the problem can be proposed if we agree that it is not necessary to require that whatever is located at the point x has to be mapped to a single point y, but that we may "spread" it around and vice versa that the necessary "quota" at y may be filled by a combination of x's.

In this case we could make a "table", $h(x, y)$ of how much of $f(x) \, dx$ goes into $g(y) \, dy$ and the "shipping plan" becomes a joint probability density $h(x, y)$, with marginals

$$f(x) = \int h(x, y) \, dy$$

$$g(y) = \int h(x, y) \, dx$$

Our minimization problem is, then, minimize

$$\int C(x - y) h(x, y) \, dx \, dy$$

among all such h.

Confronted with the problem one is of course tempted to look at the two particular cases

$$C(x - y) = |x - y| \quad \text{or} \quad |x - y|^2 .$$

The case $|x-y|$ is the original Monge problem. The quadratic case arises in many applications, particularly in fluid dynamics. Let us start by discussing the case $C(\xi) = |\xi|^2$:

The first observation or calculation that comes naturally to mind is that if $y_0(x)$ is the minimizing map, any "permutation" of k images would increase cost.

This can be better expressed in the discrete case, and we may hope it will lead us to an "Euler like" equation for this variational problem.

If π is a permutation of the x_j's it simply reads

$$\sum |y_0(x_j) - x_j|^2 \leq \sum |y_0(x_{\pi_j}) - x_j|^2$$

after some simplification

$$\sum \langle y_0(x_j), x_j \rangle \geq \sum \langle y_0(x_{\pi_j}), x_j \rangle .$$

This condition is called "cyclical" monotonicity of the map (in the case of two points x_1, x_2, it is the classical monotonicity condition of the map $\langle y(x_2) - y(x_1), x_2 - x_1 \rangle \geq 0$) and a theorem of Rockafeller asserts that the map $y(x)$ is "cyclically" monotone if and only if it is the subdifferential of a convex function $\varphi(x)$, what we would call a "convex potential" in the spirit of fluid dynamics.

To understand the meaning of $\varphi(x)$, we should go to the dual problem, that is the "shippers" point of view:

3 The dual problem:

Suppose that a shipping company wants to bid for the full transportation business. It has to charge each initiation point x_i, an amount $\mu(x_i)$ and any arrival point y_j an amount $\nu(y_j)$.

But it is constrained to charge $\mu(x_i) + \nu(y_j) \leq C(x_i - y_j)$.

If not x_i and y_j would leave the coalition and find another shipper. So the shipper wants to maximize

$$\sum \mu(x_i) + \sum \nu(y_i)$$

with the constraint that

$$\mu(x_i) + \nu(y_j) \leq C(x_i - y_j) .$$

In the continuous case, this becomes:
Maximize

$$\mathcal{K}(\mu,\nu) = \int \mu(x)f(x)\,dx + \int \nu(y)g(y)\,dy$$

with the constraint that

$$\mu(x) + \nu(y) \leq C(x - y)\ .$$

In principle, we would like to try and maximize this quantity $\mathcal{K}(\mu,\nu)$, for a reasonable family of admissible functions, say continuous.

But we note that given an admissible pair (μ,ν), we can find a better one μ, ν^*, by replacing $\nu(y)$ by

$$\nu^*(y) = \inf_x C(x - y) - \mu(x)$$

Indeed, ν^* is again admissible and $\nu^*(y) \geq \nu(y)$ for any y.

Similarly we can change μ to μ^*.

Thus the minimization process can be done in a much better class of functions, that we can call C-concave, that are of the form

$$\mu = \inf_{y \in S} C(x - y) + \nu(y)$$

(Note that if $C(x - y) = (x - y)^2$

$$-\varphi(x) = \mu(x) - |x|^2 = \inf_{y \in S} -2\langle x, y \rangle + |y|^2 - \nu(y)$$

Thus $\varphi(x)$ and $\varphi(y) = |y|^2 - \nu(y)$ are regular convex functions).

4 Existence and Uniqueness:

Since, for x, y varying in a bounded set the family of functions $C(x - y)$ is (equi) Lipschitz, it is not hard to pass to the limit and obtain a maximizing pair:

Theorem 4.1.

a) *There exists a unique maximizing pair μ_0, ν_0.*
b) *For any $x \in \Omega$, there exists at least a $y(x)$, for which*

$$C(x - y) = \mu(x) + \nu(y)\ .$$

Further, if C is strictly convex and smooth $(C^{1,\alpha})$, on can prove

Theorem 4.2.

a) *$y(x)$ is unique a.e. and the map $x \to y(x)$ is the unique optimal transportation.*

b) $y(x)$ is also defined by

$$\nabla C(x - y) = \nabla \nu(x)$$

or

$$y = x + (\nabla E)(\nabla \nu(x))$$

where (∇E) is the gradient of the Legendre transform of C, (for instance, for $C = \frac{1}{p}|x|^p$, $E = \frac{1}{q} + |x|^q$ ($\frac{1}{p} + \frac{1}{q} = 1$))

Theorem 1 is not hard, all we have to remember is that the minimization process was originally done in the space of continuous functions, that we are allowed to make just continuous perturbations of μ, ν, not necessarily C-concave ones, and hence if for some x, $\nu(x) + \mu(y) < C(x - y)$ for all $y \in \bar{\Omega}_2$, we can increase a little bit ν near x, and keep it admissible.

Theorem 2 is more delicate. The main ideas are due to Brenier:
To show that $y(x)$ is admissible, that is, according to our definition, that

$$\int \eta(y)g(y)\, dy = \int \eta(y(x))f(x)\, dx$$

corresponds basically to the Euler equation of the variational problem:
We perturb $\nu(y)$ to

$$\nu_\varepsilon = \nu(y) + \varepsilon\eta(y)$$

and $\mu(x) = \inf_y C(x - y) - \nu(y)$ to

$$\mu_\varepsilon(x) = \inf_y C(x - y) - \nu_\varepsilon(y)$$

to keep the pair μ_ε, ν_ε admissible. Thus by maximality

$$\int (\nu - \nu_\varepsilon)(y)g(y)\, dy + \int (\mu - \mu_\varepsilon)(x)f(x)\, dx = \varepsilon[\mathrm{I} + \mathrm{II}] \geq 0$$

where

$$\mathrm{I} = -\int \eta(y)g(y)\, dy$$

and

$$\mathrm{II} = \int \frac{\mu - \mu_\varepsilon}{\varepsilon}(x)f(x)\, dx$$

But by definition

$$\left|\frac{\mu - \mu_\varepsilon}{\varepsilon}\right| \leq \sup \eta < C .$$

When ε goes to zero, by dominated convergence II_ε converges to the integral of the a.e. limit of $\frac{\mu - \mu_\varepsilon}{\varepsilon}f(x)$.

But, if the $y(x)$ for which the infimum in

$$\mu(x) = \inf C(x - y) - \nu(y)$$

is attained is unique, then $\frac{\mu - \mu_\varepsilon}{\varepsilon}(x)$ converges to $\eta(y(x))$.

Once we know that $y(x)$ is admissible, if $z(x)$ is any other admissible map, we write

$$\int C\big(z(x) - x\big) f(x)\, dx \geq \int \big[\nu\big(z(x)\big) + \mu(x)\big] f(x)\, dx$$

$$= \int \nu(y)g(y)\, dy + \mu(x)f(x)\, dx$$

(since z is admissible) while

$$\int C\big(y(x) - x\big) f(x) = \int \big[\nu\big(y(x)\big) + \mu(x)\big] f(x)$$

$$= \text{(by definition of } y)$$

$$= \int \nu(y)g(y)\, dy + \int \mu(x)f(x)\, dx$$

(since y is also admissible).

5 The potential equation:

We find ourselves now in a very good position: Our mapping problem (usually a relatively hard one) has been reduced to study a single function, the potential $\mu(x)$.

The variational process made out of μ the potential of an admissible map. That is, heuristically

$$g(y) \det D\big(y(x)\big) = f(x)\ .$$

Replacing $y(x)$ by its formula

$$y = x + \nabla E(-\nabla\mu)$$

we obtain

$$g\big(x + \nabla(E(-\nabla\mu(x)))\big) \cdot \det\big(I + D(\nabla E(-\nabla\mu))\big) = f(x)$$

In the case in which

$$C(x) = \frac{1}{2}|x|^2 = E(x)$$

writing $x = \nabla(\frac{1}{2}|x|^2)$ and $\varphi(x) = \frac{1}{2}|x|^2 - \mu(x)$ the equation becomes

$$g(\nabla\varphi) \cdot \det D^2\varphi(x) = f(x)\ .$$

That is, φ is convex and it satisfies (formally) a Monge-Ampère type equation.

In the general case, always computing formally, we get

$$g(\cdots) \det\big[I + E_{ij}(-\nabla\mu)D_{ij}\mu\big] = f$$

Or if we multiply by C_{ij}, that happens to be the inverse matrix to E_{ij}, we get

$$g^* \det\big(D_{ij}\mu + C_{ij}(\nabla\mu)\big) = f$$

6 Some remarks on the structure of the equation

This last equation has a structure similar to that of the Monge-Ampère equation in a Riemannian manifold but instead of having a first order term with the structure $a_{ij}D_i\mu D_j\mu$, has the more complex one

$$E_{ij}(\nabla\mu)$$

resembling a Finsler metric term.

In confronting this equation two issues arise.

The first one, in which sense are the equations satisfied, and the second one, even if the equations are satisfied in the best of all possible senses, what can we expect of the structure of the set of solutions.

Let us answer first the second question: what can we expect about "nice" solutions of an equation of the form

$$\det\left(D^2\mu + C_{ij}(\nabla\mu)\right) = h$$

The left hand side satisfies heuristically a comparison principle (Given μ_1, μ_2 two solutions, the difference "should not" have a maximum in the interior) and it is translation invariant. Therefore, directional derivative of μ should have a "maximum principle".

Once could explore, thus, monotonicity properties of the optimal map, that is, what properties of f and g imply $y(x) \succ x$ in some order, $\succ$, in the spirit of $[C]$.

On the other hand, in the spirit of the regularity theory for fully nonlinear equations, one may ask if there are second derivative estimates. Here one should draw a parallel with solutions of divergence type equations coming from the calculus of variations. Indeed

$$\operatorname{div} D\nabla E(\nabla\mu) \tag{a}$$

very much resembles a linearization of

$$\det\left(I + \varepsilon D(\nabla E(\nabla\mu))\right) \tag{b}$$

and in general (a) possess no second derivative estimates.

So we feel that one should not expect, in general, second derivative estimates for (b) unless E is very special.

This is a very serious obstacle to a regularity theory for μ, since the linearized equation involves, as usual, second derivatives of μ in its coefficients.

About the first question, in what sense is the equation satisfied, we also have a serious difficulty. Let's consider the simple quadratic case, in which the equation is simple Monge-Ampère:

$$\det D^2\varphi\ldots$$

and φ is convex.

Since convex functions are in principle only Lipschitz, in order to make sense of $\det D^2\varphi$, the natural approach has been, precisely to look at the gradient map

$$\nabla\varphi : \Omega \to \mathbb{R}^n$$

and give an interpretation of $\det D^2$ as the Jacobian of such a map, that is the ratio between the volume of the image of a set and the volume of the set

$$\frac{|\nabla\varphi(S)|}{|S|}$$

The problem is that $\nabla\varphi(S)$ can be thought in the L^∞ sense (i.e. $\nabla\varphi \in L^\infty_{\text{loc}}$ and thus is defined a.e.) or in the maximal monotone map, i.e. $\nabla\varphi(S)$ is the set of gradients of all supporting planes to the convex function φ, on the set S.

The difference is clear in the case of $\varphi(x) = |x|$ in just one variable.

In the a.e. sense $\nabla\varphi = \pm 1$ and hence $D_{xx}\varphi = 0$. In the maximal monotone map $\nabla\varphi(0)$ is the full interval $[-1, 1]$ and thus $D_{xx}\varphi$ is the expected Dirac's δ at the origin.

Of course, this second definition is the correct one from almost any point of view, but unfortunately optimal transportation does not care much about sets of measure zero:

$$\varphi(x) = |x| + \frac{1}{2}|x|^2$$

is the optimal transportation from say $\Omega_1 = [-1, 1]$ to $\Omega_2 = [-2, -1] \cup [1, 2]$, with densities $f(y) = g(y) = 1$ but unfortunately $D^2\varphi$ has the extra density δ_0 at the origin.

Always in the particular case of $C(x) = \frac{1}{2}|x|^2$, the natural geometric condition to impose on Ω_2 to avoid this difficulty is simple: Ω_2 must be convex.

Indeed it is easy to see that the difference between $\det D^2\varphi$ (a.e.) and $\det D^2\varphi$ (max. mon.) consists of a singular measure, whose image by the gradient map is always contained in the (closure of) the convex envelope of the image of the regular part.

Thus, if Ω_2 is convex, and since we have the compatibility condition that $\int_{\Omega_2} f(x) = \int_{\Omega_2} g(y)$, any extra singular measure has nowhere to go inside Ω_2.

Once we know that φ satisfies the equation in the maximal monotone sense (Alexandrov sense) there exists a reasonable local regularity theory that asserts that the map $\nabla\varphi$ is, as expected, "one derivative better" than $f(x)$, $g(y)$.

For general cost functions is is not clear how to proceed both to find the appropriate condition on Ω_2, and how to develop at least the first steps on a local regularity theory, asserting for instance that the map is continuous.

Part of the difficulty is that convexity is simultaneously a local (infinitesimal) and global condition, that is $D^2\varphi \geq 0$ or the graph of φ stays below the segment joining two points, while in the general case that does not seem to be the case (with $-C(x - y)$ replacing linear functions and $-[D^2\mu + E_{ij}(\nabla\mu)]$ replacing $D^2\varphi$.

This becomes evident for instance with the issue of $\nabla\mu$ (a.e.) versus $\nabla\mu$ (maximal monotone).

Since $\mu = \inf_y C(x - y) - \nu(y)$ one may suggest that $\nabla\mu$ (maximal monotone) should be the $\nabla\mu(x)$ coming from all supporting cost functions $C(x-y)$ at the point x. But a cost function $C(x - y)$ that supports μ locally near x does not necessarily support it globally.

A possible way out is to study these cases for which both $f(x)$, $g(y)$ do not vanish, thus $\Omega_1 = \Omega_2 = \mathbb{R}^n$ or periodic problems ($f(x)$, $g(y)$ periodic) where hopefully, whatever singular part one should add to the a.e. definition of the map, has nowhere to go again.

Finally, some remarks on the case $C(x - y) = |x - y|$, the classic Monge problem.

In this case, strict convexity is lost and therefore ∇C is not anymore a "nice" invertible map from $\mathbb{R}^n$ to $\mathbb{R}^n$.

So if one tries to reconstruct the map from μ, ν the solution pair of the dual problem, $\nabla\mu$ only gives us the general direction of the optimal transportation, i.e.

$$\nabla\mu(x) = \frac{y - x}{|y - x|}$$

but not the distance.

In fact, the optimal allocation is, in general, not unique: If we want to transport, in one dimension the segment $[-2, -1] = I_0$ onto $[1, 2] = I_1$, the cost function $|y-x|$ becomes simply $y-x$, and from the change of variable formula, we see that any measure preserving transformation $y(x)$ is a minimizer.

So, for instance we can split I_0, I_1 into a bunch of little intervals and map each other more or less arbitrarily and still have a minimizer.

On the other hand, in more than one dimension, optimal maps have a strong geometric restriction: If $x_1 \rightarrow x_2$ and $y_1 \rightarrow y_2$, the segments $[x_1, y_1]$, $[x_2, y_2]$ cannot intersect. (Remember the discrete case.)

Further, if x_2 lies in the interior of $[x_1, y_1]$, y_2 must be aligned to $[x_1, y_1]$. Otherwise transportation can be improved.

Thus, transportation is aligned along "transportation rays". That is, if x_1 goes to y_1, all points in $\Omega_1 \cap [x_1, y_1]$ are mapped to the line containing $[x_1, y_1]$.

This is easy to visualize if for instance Ω_1 is contained in $\{x : x_n < 0\} = \mathbb{R}^n_-$, and Ω_2 in $\{x : x_n > 0\} = \mathbb{R}^n_+$.

Then transportation occurs along rays going from left to right.

There are several geometrical quantities one can look at to try to understand what is going on.

For instance, if Ω_1 is a long vertical rectangle and Ω_2 a horizontal one, one can see that in general, we cannot expect a "nice, clean" foliation of Ω_1, Ω_2 by transport rays, and that the domains must split in patches each one foliated by these "transportation" rays.

Another observation, from the definition of μ, ν as

$$\mu(x) = \inf_{y \in S} C(x - y) - \nu(y)$$

we see that if we define the two auxiliary functions

$$h(z) = \inf_{y \in \Omega_2} |z - y| - \nu(y)$$

and

$$g(z) = \sup_{x \in \Omega_1} \mu(x) - |z - x|$$

then $h(z) = g(z)$ along transportation rays, where both are linear.

Since along $x_n = 0$, h is $C^{1,1}$ by above (quasiconcave) and g $C^{1,1}$ by below (quasiconvex). The direction $\mathcal{T}(x)$ of transportation rays is Lipschitz along $\{x_n = 0\}$.

Two possible ideas to construct a solution are then to write an equation for the potential and the infinitesimal transportation along these rays, given by the mass balance (this is the approach of Evans-Gangbo) or to pass to the limit on a strictly ε-approximation of the limiting problem.

This last approach was worked out by Trudinger and Wang, and Feldman, McCann and the author, and was recently completed by L. Ambrosio and A. Pratelli by incorporating higher order T-convergence arguments to obtain strong convergence of the map.

Bibliographical Remarks: There is extensive work in this area.

Instead of attempting a long list of references let me mention a few names whose work is addressed. General issues of existence and regularity, that can be easily traced in the MathSciNet: L. Ambrosio, Y. Bremier, L.C. Evans, M. Feldman, W. Gangbo, R. McCann, N. Trudinger, J. Urbas, X.J. Wang.

Optimal Shapes and Masses, and Optimal Transportation Problems

Giuseppe Buttazzo[1] and Luigi De Pascale[2]

[1] Dipartimento di Matematica, Università di Pisa
   Via Buonarroti 2, 56127 Pisa, ITALY
   buttazzo@dm.unipi.it
[2] Dipartimento di Matematica Applicata, Università di Pisa
   Via Bonanno Pisano 25/b, 56126 Pisa, ITALY
   depascal@dm.unipi.it

Summary. We present here a survey on some shape optimization problems that received particular attention in the last years. In particular, we discuss a class of problems, that we call mass optimization problems, where one wants to find the distribution of a given amount of mass which optimizes a given cost functional. The relation with mass transportation problems will be discussed, and several open problems will be presented.

Keywords. Shape optimization, mass transportation, Monge-Kantorovich equation.
MSC 2000. 49Q10, 49J45, 74P05.

1 Introduction

The purpose of these notes is to give a survey on some problems in shape and mass optimization that received a lot of attention in the mathematical literature in the recent years. After a presentation of shape optimization problems in a quite general framework we give some examples that nowadays can be considered classic.

A shape optimization problem is a minimization problem where the unknown variable runs over a class of domains; then every shape optimization problem can be written in the form

$$\min \left\{ F(A) \ : \ A \in \mathcal{A} \right\} \tag{1}$$

where $\mathcal{A}$ is the class of admissible domains and F is the cost function that one has to minimize over $\mathcal{A}$. It has to be noticed that the class $\mathcal{A}$ of admissible domains does not have any linear or convex structure, so in shape optimization problems it is meaningless to speak of convex functionals and similar notions. Moreover, even if several topologies on families of domains are available, in general there is not an a priori choice of a topology in order to apply the direct methods of the calculus of variations for obtaining the existence of at least an optimal domain.

We shall not give here a detailed presentation of the many problems and results in this very wide field, but we limit ourselves to discuss some model problems. We refer the reader interested in a deeper knowledge and analysis of this fascinating field to one of the several books on the subject ([3], [114], [142], [146]), to the notes by L. Tartar [149], or to the recent collection of lecture notes by D. Bucur and G. Buttazzo [37].

In many situations, a shape (or also a mass) optimization problem can be seen as an optimal control problem, where the state variable solves a PDE of elliptic type, and the control variable is given by the unknown domain (or mass distribution). We want to stress that, as it also happens in other kinds of optimal control problems, in several situations an optimal domain does not exist; this is mainly due to the fact that in these cases the minimizing sequences are highly oscillating and converge to a limit object only in a *"relaxed"* sense. Then we may have, in these cases, only the existence of a *"relaxed solution"*, suitably defined, that in general is not a domain, and whose characterization may change from problem to problem.

A general procedure to relax optimal control problems can be successfully developed (see for instance [18], [60]) by using the Γ-convergence scheme which provides the right topology to use for sequences of admissible controls. In particular, for shape optimization problems, this provides the right notion of convergence for sequences of domains. However, if in the minimization problem (1), either the class $\mathcal{A}$ of admissible domains or the cost functional F constrain the admissible domains to fulfill some sufficiently strong geometrical conditions, then the existence of an optimal domain may be obtained. We shall see some examples where these situations occur.

As it happens in all optimization problems, the qualitative description of the optimal solutions of a shape optimization problems, is usually obtained by means of the so called *necessary conditions of optimality*. These conditions have to be derived from the comparison of the cost of an optimal solution A_{opt} with the cost of other suitable admissible choices, close enough to A_{opt}. This procedure is what is usually called a *variation* near the solution. We want to stress that in shape and mass optimization problems, the notion of neighbourhood is not always a priori clear; the possibility of choosing a domain variation could then be rather wide, and this often provides several necessary conditions of optimality.

In general, explicit computations of optimal shapes and masses are difficult to obtain, and one should develop efficient numerical schemes to produce approximated solutions; we will not enter this important field and we refer the interested reader to some recent books and papers (see references).

The strict relation between some mass optimization problems and optimal transportation results has been recently shown by Bouchitté and Buttazzo in [23]; we shall shortly summarize the results obtained and we shall present some new challenging problems.

2 Some classical problems

In this section we present some classical examples of shape optimization problems that can be written in the form (1).

2.1 The isoperimetric problem

The isoperimetric problem is certainly one of the oldest shape optimization problems; given a closed set $Q \subset \mathbb{R}^N$, the constraint set, it consists in finding, among all Borel subsets $A \subset Q$, the one which minimizes the perimeter, once its Lebesgue measure, or more generally the quantity $\int_A f(x)\, dx$ for a given function $f \in L^1_{loc}(\mathbb{R}^N)$, is prescribed. With this notation the isoperimetric problem can be then formulated in the form (1) if we take

$$F(A) = \text{Per}(A),$$
$$\mathcal{A} = \Big\{ A \subset Q \ : \ \int_A f(x)\, dx = c \Big\}.$$

Here the perimeter of a Borel set A is defined as usual by

$$\text{Per}(A) = \int |D1_A| = \mathcal{H}^{N-1}(\partial^* A)$$

where $D1_A$ is the distributional derivative of the characteristic function of A, $\mathcal{H}^{N-1}$ is the $N-1$ dimensional Hausdorff measure, and $\partial^* A$ is the reduced boundary of A in the sense of geometric measure theory. By using the properties of the BV spaces, when Q is bounded we obtain the lower semicontinuity and the coercivity of the perimeter for the L^1 convergence; this enables us to apply the direct methods of the calculus of variations and to obtain straightforward the existence of an optimal solution for the problem

$$\min \Big\{ \text{Per}(A) \ : \ A \subset Q, \int_A f\, dx = c \Big\}. \tag{2}$$

It is also very simple to show that in general the problem above may have no solution if we drop the assumption that Q is bounded (see for instance [16],[37]). On the other hand, it is very well known that the classical isoperimetric problem, with $Q = \mathbb{R}^N$ and $f \equiv 1$, admits a solution which is any ball of measure c, even if the complete proof of this fact requires very delicate tools, especially when the dimension N is larger than 2. A complete characterization of pairs (Q, f) which provide the existence of a solution for the problem (2) seems to be difficult.

Assume now that a set A be a solution of problem (2) and that A is regular enough to perform all the operations we need. Therefore we have that, at least locally, the boundary ∂A can be represented by the graph of a function $u(x)$, where x varies in a small neighbourhood ω. By performing now the usual first variation argument with the functional

$$\int_\omega \sqrt{1 + |\nabla u|^2}\, dx$$

we obtain that the function u must satisfy the partial differential equation

$$-\mathrm{div}\Big(\frac{\nabla u}{\sqrt{1 + |\nabla u|^2}}\Big) = \text{constant} \qquad \text{in } \omega.$$

The term $-\mathrm{div}\big(\nabla u/\sqrt{1 + |\nabla u|^2}\big)$ represents the mean curvature of ∂A written in Cartesian coordinates; therefore we have found for a regular solution A of the isoperimetric problem (2) the following necessary condition of optimality:

the mean curvature of ∂A is constant in the interior points of Q.

Actually, the regularity of ∂A does not need to be assumed as a hypothesis but is a consequence of some suitable conditions on the datum f (regularity theory).

A variant of the isoperimetric problem consists in not counting some parts of the boundary ∂A in the cost functional. More precisely, if Q is the closure of an open set Ω with a Lipschitz boundary, we may consider problem (2) with $\mathrm{Per}(A)$ replaced by the cost functional

$$\mathrm{Per}_\Omega(A) = \int_\Omega |D1_A| = \mathcal{H}^{N-1}(\Omega \cap \partial^* A).$$

The existence of a solution when Ω is bounded still holds, as above, together with nonexistence examples when this boundedness condition is dropped.

2.2 The Newton's problem of optimal aerodynamical profiles

The problem of finding the best aerodynamical profile for a body in a fluid stream under some constraints on its size is another classical question which can be considered as a shape optimization problem. This problem was first considered by Newton, who gave a rather simple variational expression for the aerodynamical resistance of a convex body in a fluid stream, assuming that the competing bodies are radially symmetric, which makes the problem onedimensional. Here are his words (from *Principia Mathematica*):

If in a rare medium, consisting of equal particles freely disposed at equal distances from each other, a globe and a cylinder described on equal diameter move with equal velocities in the direction of the axis of the cylinder, (then) the resistance of the globe will be half as great as that of the cylinder. ... I reckon that this proposition will be not without application in the building of ships.

Under the assumption that the resistance is due to the impact of fluid particles against the body surface, that the particles are supposed all independent (which is quite reasonable if the fluid is rarefied), and that the tangential friction is negligible, by simple geometric considerations we may obtain the following expression of the resistance along the direction of the fluid stream, where we normalize all the physical constants to one:

$$F(u) = \int_\Omega \frac{1}{1 + |Du|^2} \, dx. \tag{3}$$

In the expression above we denoted by Ω the cross section of the body at the basis level, and by $u(x)$ a function whose graph is the body boundary. The geometrical constraint in the problem consists in requiring that the admissible competing bodies be convex; this is also consistent with the physical assumption that all the fluid particles hit the body at most once. In problem (3) this turns out to be equivalent to assume that Ω is convex and that $u : \Omega \to [0, +\infty[$ is concave.

The problem, as considered by Newton, is when Ω is a disc (of radius R) and the competing functions are supposed a priori with a radial symmetry. In this case, after integration in polar coordinates, the expression of the resistance takes the form

$$F(u) = 2\pi \int_0^R \frac{r}{1 + |u'(r)|^2} \, dr$$

so that the problem of the determination of the optimal radial profile becomes

$$\min \left\{ \int_0^R \frac{r}{1 + |u'(r)|^2} \, dr \; : \; u \text{ concave}, \, 0 \le u \le M \right\}. \tag{4}$$

If instead of the functions $u(r)$ we use the functions $v(t) = u^{-1}(M - t)$, the minimization problem (4) can be rewritten in the more traditional form

$$\min \left\{ \int_0^M \frac{vv'^3}{1 + v'^2} \, dt \; : \; v \text{ concave}, \, v(0) = 0, \, v(M) = R \right\}. \tag{5}$$

The functional appearing in (5) has to be intended in the sense of BV functions; in fact v is a nondecreasing function, so that v' is a nonnegative measure and the precise expression of the functional in (5) is

$$\int_0^M \frac{vv'^3_a}{1 + v'^2_a} \, dt + \int_{[0,M]} vv'_s = \frac{R^2}{2} - \int_0^M \frac{vv'_a}{1 + v'^2_a} \, dt$$

where v'_a and v'_s are respectively the absolutely continuous and singular parts of the measure v' with respect to Lebesgue measure.

It is possible to compute explicitly the solution of the minimization problem (4) by integrating its Euler-Lagrange equation which reads as

$$ru' = C(1 + u'^2)^2 \qquad \text{on } \{u' \neq 0\}$$

for a suitable constant $C < 0$. By introducing the function

$$f(t) = \frac{t}{(1+t^2)^2}\left(-\frac{7}{4} + \frac{3}{4}t^4 + t^2 - \log t\right) \qquad \forall t \geq 1$$

and the quantities

$$T = f^{-1}(M/R), \qquad r_0 = \frac{4RT}{(1+T^2)^2},$$

the solution u can be obtained as

$$u(r) = M \qquad \forall r \in [0, r_0],$$

and for $r > r_0$ its expression is given in parametric form by:

$$\begin{cases} r(t) = \dfrac{r_0}{4t}(1+t^2)^2 \\ u(t) = M - \dfrac{r_0}{4}\left(-\dfrac{7}{4} + \dfrac{3}{4}t^4 + t^2 - \log t\right) \end{cases} \qquad \forall t \in [1, T].$$

We have $|u'(r)| > 1$ for all $r > r_0$ and $|u'(r_0^+)| = 1$; moreover, the optimal radial solution can be shown to be unique.

The optimal radial solution of the Newton problem for $M = R$ is shown in figure below.

Coming back to the Newton problem in its general form (not necessarily restricted to radial functions)

$$\min\left\{F(u) \; : \; u \text{ concave}, \, 0 \leq u \leq M\right\}, \tag{6}$$

where the functional F is given by (3), we notice that the integral functional F above is neither convex nor coercive. Therefore, the usual direct methods of the calculus of variations for obtaining the existence of an optimal solution may fail. However, due to the concavity constraint, the existence of a minimizer u still holds, as it has been proved in [59]. A complete discussion on the problem above can be found in [37] where all the concerning references are quoted. Here we simply recall an interesting necessary condition of optimality (see [119]): it turns out that on every open set ω where u is of class C^2 we obtain

$$\det D^2 u(x) = 0 \qquad \forall x \in \omega.$$

In particular, as it is easy to see, this excludes that in the case $\Omega = B(0, R)$ the solution u be radially symmetric. A profile better than all radial profiles with the same height has been found by Guasoni in [111] and is graphically represented in the figure below. Nevertheless, a complete characterization of the optimal solutions in the case $\Omega = B(0, R)$ is not yet know.

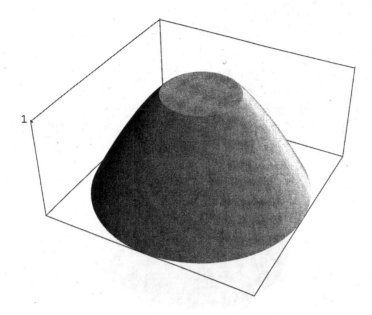

Fig. 1. The optimal radial shape for $M = R$.

2.3 Optimal Dirichlet regions

We consider now the model example of a Dirichlet problem over an unknown domain, which has to be optimized according to a given cost functional. Denoting by f a given function, say in $L^2(\mathbb{R}^N)$, by A the unknown domain, and by u_A the solution of the elliptic equation

$$-\Delta u = f \text{ in } A, \qquad u \in H_0^1(A)$$

extended by zero outside A, the optimization problem we consider is

$$\min\left\{ \int_Q j(x, u_A)\, dx \ : \ A \subset Q\right\}.$$

Here Q is a given bounded domain of $\mathbb{R}^N$ and $j(x, s)$ a given integrand.

We refer to the lecture notes [37] for a more complete discussion on this topic and for a wide list of references devoted to this subject; here we want only to summarize the different situations that may occur.

It is well known that in general one should not expect the existence of an optimal solution; for instance, if we take

$$j(x, s) = |s - a(x)|^2$$

in general, even if the function $a(x)$ is constant, the existence of an optimal solution may fail. However, the existence of an optimal domain occurs in some particular cases that we list below.

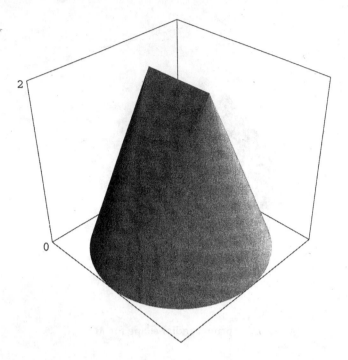

Fig. 2. A nonradial profile better than all radial ones.

i) On the class of admissible domains some rather severe geometrical constraints are imposed. For instance the so called it exterior cone condition is sufficient to imply the existence of an optimal solution. It consists in requiring the existence of a given height h and opening ω such that for all domains A in the admissible class $\mathcal{A}$ and for all points $x_0 \in \partial A$ a cone with vertex x_0, height h and opening ω is contained in the complement of A (see [48], [49], [50]).

ii) The cost functional fulfills some particular qualitative assumptions. In particular, if a functional $F(A)$ is nonincreasing with respect to the set inclusion (and satisfies a rather mild lower semicontinuity assumption with respect to a γ-convergence on domains, suitably defined), then the minimization problem

$$\min \big\{ F(A) \ : \ A \subset Q, \ \operatorname{meas}(A) \leq m \big\}$$

admits a solution (see [55]).

iii) The problem is of a very special type, involving only the first two eigenvalues of the Laplace operator, where neither geometrical constraints nor monotonicity of the cost are required (see [38]).

We want to stress that, though quite particular, the previous case ii) covers some interesting situations. For instance cost functionals of the form

$$F(A) = \int_Q j(x, u_A)\, dx$$

with $j(x, \cdot)$ nonincreasing, fulfill the required assumptions. Another interesting situation is given by functions of eigenvalues $\lambda_k(A)$ of an elliptic operator L with Dirichlet boundary conditions on ∂A. In this case, setting $\Lambda(A) = (\lambda_k)_{k \in \mathbb{N}}$, and using the well known fact that the eigenvalues of an elliptic operator are nonincreasing functions of the domain, the functional

$$F(A) = \Phi\big(\Lambda(A)\big)$$

fulfills the required assumptions as soon as the function $\Phi : \mathbb{R}^{\mathbb{N}} \to [0, +\infty]$ is lower semicontinuous, that is

$$(\Lambda_n)_k \to (\Lambda)_k \quad \forall k \in \mathbb{N} \quad \Rightarrow \quad \Phi(\Lambda) \leq \liminf_n \Phi(\Lambda_n),$$

and nondecreasing, that is

$$(\Lambda_1)_k \leq (\Lambda_2)_k \quad \forall k \in \mathbb{N} \quad \Rightarrow \quad \Phi(\Lambda_1) \leq \Phi(\Lambda_2).$$

2.4 Optimal mixtures of two conductors

An interesting question that can be seen in the form of a shape optimization problem is the determination of the optimal distribution of two given conductors (for instance in the thermostatic model, where the state function is the temperature of the system) into a given set. If Ω denotes a given bounded open subset of $\mathbb{R}^N$ (the prescribed container), denoting by α and β the conductivities of the two materials, the problem consists in filling Ω with the two materials in the most performant way according to some given cost functional. The volume of each material can also be prescribed. We denote by A the domain where the conductivity is α and by $a_A(x)$ the conductivity coefficient

$$a_A(x) = \alpha 1_A(x) + \beta 1_{\Omega \setminus A}(x).$$

Then the state equation which associates the control A to the state u (the temperature of the system, once the conductor α fills the domain A) becomes

$$\begin{cases} -\operatorname{div}\big(a_A(x)Du\big) = f \text{ in } \Omega \\ u = 0 \qquad\qquad\quad \text{ on } \partial\Omega, \end{cases} \tag{7}$$

where f is the (given) source density. We denote by u_A the unique solution of (7).

It is well known (see for instance Kohn and Strang [117], Murat and Tartar [138]) that if we take as a cost functional an integral of the form

$$\int_\Omega j(x, 1_A, u_A, Du_A)\, dx$$

in general an optimal configuration does not exist. However, the addition of a perimeter penalization is enough to imply the existence of classical optimizers. More precisely, we take as a cost the functional

$$J(u, A) = \int_\Omega j(x, 1_A, u, Du)\, dx + \sigma \mathrm{Per}_\Omega(A)$$

where $\sigma > 0$, and the optimal control problem then takes the form

$$\min \big\{ J(u, A) \ : \ A \subset \Omega,\ u \text{ solves } (7) \big\}. \tag{8}$$

A volume constraint of the form meas$(A) = m$ could also be present. The proof of the existence of an optimal classical solution for problem (8) uses the following result.

Proposition 2.1. *For every $n \in \mathbb{N}$ let $a_n(x)$ be a $N \times N$ symmetric matrix, with measurable coefficients, such that the uniform ellipticity condition*

$$c_0|z|^2 \le a_n(x)z \cdot z \le c_1|z|^2 \qquad \forall x \in \Omega,\ \forall z \in \mathbb{R}^N \tag{9}$$

holds with $0 < c_0 \le c_1$ (independent of n). Given $f \in H^{-1}(\Omega)$ denote by u_n the unique solution of the problem

$$-\mathrm{div}\big(a_n(x)Du\big) = f, \qquad u_n \in H_0^1(\Omega). \tag{10}$$

If $a_n(x) \to a(x)$ a.e. in Ω then $u_n \to u$ weakly in $H_0^1(\Omega)$, where u is the solution of (10) with a_n replaced by a.

Proof. By the uniform ellipticity condition (9) we have

$$c_0 \int_\Omega |Du_n|^2\, dx \le \int_\Omega f u_n\, dx$$

and, by the Poincaré inequality we have that u_n are bounded in $H_0^1(\Omega)$ so that a subsequence (still denoted by the same indices) converges weakly in $H_0^1(\Omega)$ to some v. All we have to show is that $v = u$ or equivalently that

$$-\mathrm{div}\big(a(x)Dv\big) = f. \tag{11}$$

This means that for every smooth test function ϕ we have

$$\int_\Omega a(x)DvD\phi\, dx = \langle f, \phi \rangle.$$

Then it is enough to show that for every smooth test function ϕ we have

$$\lim_{n \to +\infty} \int_\Omega a_n(x)Du_nD\phi\, dx = \int_\Omega a(x)DvD\phi\, dx.$$

This is an immediate consequence of the fact that ϕ is smooth, $Du_n \rightarrow Dv$ weakly in $L^2(\Omega)$, and $a_n \rightarrow a$ a.e. in Ω remaining bounded.

Another way to show that (11) holds is to verify that v minimizes the functional

$$F(w) = \int_\Omega a(x)DwDw\,dx - 2\langle f, w \rangle \qquad w \in H_0^1(\Omega). \qquad (12)$$

Since the function $\alpha(s, z) = sz \cdot z$, defined for $z \in \mathbb{R}^N$ and for s symmetric positive definite $N \times N$ matrix, is convex in z and lower semicontinuous in s, the functional

$$\Phi(a, \xi) = \int_\Omega a(x)\xi \cdot \xi\,dx$$

is sequentially lower semicontinuous with respect to the strong L^1 convergence on a and the weak L^1 convergence on ξ (see for instance [52]). Therefore we have

$$F(v) = \Phi(a, Dv) - 2\langle f, v \rangle \leq \liminf_{n \to +\infty} \Phi(a_n, Du_n) - 2\langle f, u_n \rangle = \liminf_{n \to +\infty} F(u_n).$$

Since u_n minimizes the functional F_n defined as in (12) with a replaced by a_n, we also have for every $w \in H_0^1(\Omega)$

$$F_n(u_n) \leq F_n(w) = \int_\Omega a_n(x)DwDw\,dx - 2\langle f, w \rangle$$

so that taking the limit as $n \to +\infty$ and using the convergence $a_n \to a$ we obtain

$$\liminf_{n \to +\infty} F_n(u_n) \leq \int_\Omega a(x)DwDw\,dx - 2\langle f, w \rangle = F(w).$$

Thus $F(v) \leq F(w)$ which shows what required.

Remark 2.1. The result of proposition above can be equivalently rephrased in terms of G-convergence by saying that for uniformly elliptic operators of the form $-\text{div}(a(x)Du)$ the G-convergence is weaker than the L^1-convergence of coefficients. Analogously, we can say that the functionals

$$G_n(w) = \int_\Omega a_n(x)DwDw\,dx$$

Γ-converge to the functional G defined in the same way with a in the place of a_n.

Corollary 2.1. *If $A_n \to A$ in $L^1(\Omega)$ then $u_{A_n} \to u_A$ weakly in $H_0^1(\Omega)$.*

A more careful inspection of the proof of Proposition 2.1 shows that the following stronger result holds.

Proposition 2.2. *Under the same assumptions of Proposition 2.1 the convergence of u_n is actually strong in $H_0^1(\Omega)$.*

Proof. In Proposition 2.1 we have already seen that $u_n \to u$ weakly in $H_0^1(\Omega)$, which gives $Du_n \to Du$ weakly in $L^2(\Omega)$. Denoting by $c_n(x)$ and $c(x)$ the square root matrices of $a_n(x)$ and $a(x)$ respectively, we have that $c_n \to c$ a.e. in Ω remaining equi-bounded. Then $c_n(x)Du_n$ converge to $c(x)Du$ weakly in $L^2(\Omega)$. Multiplying equation (10) by u_n and integrating by parts we obtain

$$\int_\Omega a(x)DuDu\, dx = \langle f, u \rangle = \lim_{n \to +\infty} \langle f, u_n \rangle = \lim_{n \to +\infty} \int_\Omega a_n(x)Du_nDu_n\, dx.$$

This implies that

$$c_n(x)Du_n \to c(x)Du \qquad \text{strongly in } L^2(\Omega).$$

Multiplying now by $\big(c_n(x)\big)^{-1}$ we finally obtain the strong convergence of Du_n to Du in $L^2(\Omega)$.

We are now in a position to obtain an existence result for the optimization problem (8). On the function j we only assume that it is nonnegative, Borel measurable, and such that $j(x, s, z, w)$ is lower semicontinuous in (s, z, w) for a.e. $x \in \Omega$.

Theorem 2.1. *Under the assumptions above the minimum problem (8) admits at least a solution.*

Proof. Let (A_n) be a minimizing sequence; then $\mathrm{Per}_\Omega(A_n)$ are bounded, so that, up to extracting subsequences, we may assume (A_n) is strongly convergent in the L^1 sense to some set $A \subset \Omega$. We claim that A is a solution of problem (8). Let us denote by u_n a solution of problem (7) associated to A_n; by Proposition 2.2 (u_n) converges strongly in $H_0^1(\Omega)$ to some $u \in H_0^1(\Omega)$. Then by the lower semicontinuity of the perimeter and by Fatou's lemma we have

$$J(u, A) \le \liminf_{n \to +\infty} J(u_n, A_n)$$

which proves the optimality of A.

Remark 2.2. The existence result above still holds, with the same proof, when volume constraints of the form $\mathrm{meas}(A) = m$ are present. Indeed this constraint passes to the limit when $A_n \to A$ strongly in $L^1(\Omega)$.

The existence result above shows the existence of a classical solution for the optimization problem (8). This solution is simply a set with finite perimeter and additional assumptions have to be made in order to prove further regularity. For instance in [8] Ambrosio and Buttazzo considered the similar problem

$$\min \left\{ E(u, A) + \sigma \mathrm{Per}_\Omega(A) \ : \ u \in H_0^1(\Omega), \ A \subset \Omega \right\}$$

where $\sigma > 0$ and

$$E(u, A) = \int_\Omega \left[a_A(x)|Du|^2 + 1_A(x)g_1(x, u) + 1_{\Omega \setminus A}g_2(x, u) \right] dx.$$

They showed that every solution A is actually an open set provided g_1 and g_2 are Borel measurable and satisfy the inequalities

$$g_i(x, s) \geq \gamma(x) - k|s|^2 \qquad i = 1, 2$$

where $\gamma \in L^1(\Omega)$ and $k < \alpha \lambda_1$, being λ_1 the first eigenvalue of $-\Delta$ on Ω.

3 Mass optimization problems

In this section we present an optimization problem that we call *mass optimization problem*; it plays a central role in many questions in Applied Mathematics and Engineering. It consists in finding the elastic structure (seen as a distribution of a given amount of elastic material) that, for a given system f of loads and for a given total mass, gives the best resistance in terms of minimal compliance. The unknown mass distribution is then a nonnegative measure which may vary in the class of admissible choices, with total mass prescribed, and support possibly constrained in a given *design region*.

In order to take into account also forces which may concentrate on lower dimensional sets we consider a force field $f \in \mathcal{M}(\mathbb{R}^N; \mathbb{R}^N)$, the class of all $\mathbb{R}^N$-valued measures on $\mathbb{R}^N$ with finite total variation and with compact support. The class of smooth displacements we consider is the Schwartz space $\mathcal{D}(\mathbb{R}^N; \mathbb{R}^N)$ of C^∞ functions with compact support; similarly, the notation $\mathcal{D}'(\mathbb{R}^N; \mathbb{R}^N)$ stands for the space of vector valued distributions and, for a given nonnegative measure μ, $L_\mu^p(\mathbb{R}^N; \mathbb{R}^d)$ denotes the space of p-integrable functions with respect to μ with values in $\mathbb{R}^d$.

For every $N \times N$ matrix z we denote by z^{sym} the symmetric part of z and by $e(u)$ the strain tensor $(Du)^{sym}$; in this way, for a given smooth displacement $u : \mathbb{R}^N \to \mathbb{R}^N$ we denote by $j(Du) = j(e(u))$ the stored elastic energy density associated to u, where

$$j(z) = \beta|z^{sym}|^2 + \frac{\alpha}{2}|tr(z^{sym})|^2 \tag{13}$$

being α and β are the so called Lamé constants. This is the case when the material to distribute is a homogeneous isotropic linearly elastic material; the same analysis holds if more generally we assume:

i) j is convex and positively p-homogeneous, with $p > 1$;

ii) $j(z) = j(z^{sym})$;

iii) there exist two positive constants α_1 and α_2 such that

$$\alpha_1|z^{sym}|^p \leq j(z) \leq \alpha_2|z^{sym}|^p \qquad \forall z \in \mathbb{R}^{N \times N}.$$

For a given mass distribution μ the stored elastic energy of a smooth displacement $u \in \mathcal{D}(\mathbb{R}^N; \mathbb{R}^N)$ is given by

$$J(\mu, u) = \int j(Du)\, d\mu$$

so that the total energy associated to μ and relative to a smooth displacement u is

$$E(\mu, u) = J(\mu, u) - \langle f, u \rangle$$

where $\langle f, u \rangle$ represents the work of the force field f.

In order to take into account possibly prescribed Dirichlet boundary conditions, we consider a closed subset Σ of $\mathbb{R}^N$ (when $\Sigma = \emptyset$ the problem is called of *pure traction* type) and we impose the admissible displacements vanish on Σ. Thus we may now define the energy of a measure μ as the infimum

$$\mathcal{E}(\mu) = \inf \left\{ E(\mu, u) \; : \; u \in \mathcal{D}(\mathbb{R}^N; \mathbb{R}^N), \; u = 0 \text{ on } \Sigma \right\} \qquad (14)$$

and the compliance $\mathcal{C}(\mu)$ is then defined as

$$\mathcal{C}(\mu) = -\mathcal{E}(\mu).$$

The optimization problem we want to consider is then

$$\min \left\{ \mathcal{C}(\mu) \; : \; \mu \in \mathcal{M}^+(\mathbb{R}^N), \; \int d\mu = m, \; \mathrm{spt}\mu \subset K \right\} \qquad (15)$$

where the total mass constraint $\int d\mu = m$ is present, and where a *design region* constraint is also possible, which turns out to give a closed subset K of $\mathbb{R}^N$ and to limit the analysis only to mass distributions which vanish outside K. We assume that K is the closure $\overline{\Omega}$ of a smooth connected bounded open subset Ω of $\mathbb{R}^N$. It should also be noticed that the problem above is a variational model which describes the behaviour of *light structures*, where the force due to their own weight is neglected.

Since for a fixed smooth admissible displacement u the mapping $\mu \mapsto E(\mu, u)$ is affine and continuous for the weak* convergence, the functional $\mathcal{C}(\mu)$ turns out to be convex and lower semicontinuous. Therefore, by direct methods of the calculus of variations we obtain the following existence result.

Theorem 3.1. *The mass optimization problem (15) admits at least a solution.*

Remark 3.1. The same formulation can be used for the case of scalar state functions u; in this case the prototype of the function $j(z)$ is the Dirichlet energy density

$$j(z) = \frac{1}{2}|z|^2$$

and the optimization problem (15) describes the optimal distribution of a given amount of conducting (for instance in the thermostatic model) material, being f the heat sources density.

Remark 3.2. We want to stress that we may have $C(\mu) = +\infty$ for some measures μ; this happens for instance in the case when the force field f concentrates on sets of dimension smaller than $n - 1$ and the mass distribution μ is the Lebesgue measure. However, these "singular" measures μ which have infinite compliance are ruled out from our discussion because we look for the minimization of the compliance functional $C(\mu)$.

By standard duality arguments (see for instance [98]) we may rewrite the compliance $C(\mu)$ in the form

$$C(\mu) = \inf \left\{ \int j^*(\sigma) \, d\mu \; : \; \sigma \in L^{p'}_\mu(\mathbb{R}^N; \mathbb{R}^{N \times N}), \right.$$

$$\left. - \operatorname{div}(\sigma\mu) = f \text{ in } \mathcal{D}'(\mathbb{R}^N \setminus \Sigma; \mathbb{R}^N) \right\} \quad (16)$$

and the infimum in (16) is actually a minimum as soon as $C(\mu)$ is finite.

In order to characterize the optimal solutions μ_{opt} of problem (15) by means of necessary and sufficient conditions of optimality it is convenient to introduce the quantity

$$I(f, \Sigma, \Omega) = \sup \left\{ \langle f, u \rangle \; : \; u = 0 \text{ on } \Sigma, \; j(Du) \leq 1/p \text{ on } \overline{\Omega} \right\}. \quad (17)$$

The following result holds (we refer to [23] for the proof).

Proposition 3.1. *The mass optimization problem (15) is nontrivial, in the sense that the compliance functional C is not identically $+\infty$, provided $I(f, \Sigma, \Omega)$ is finite. Moreover, for every nonnegative measure μ with $\int d\mu = m$ and $\operatorname{spt}\mu \subset \overline{\Omega}$ we have*

$$C(\mu) \geq \frac{(I(f, \Sigma, K))^{p'}}{p' m^{1/(p-1)}}. \quad (18)$$

Finally, there exists a nonnegative measure μ with $\int d\mu = m$ and $\operatorname{spt}\mu \subset K$ such that

$$C(\mu) \leq \frac{(I(f, \Sigma, \Omega))^{p'}}{p' m^{1/(p-1)}}.$$

The optimal mass distribution μ_{opt} then verifies the equality

$$C(\mu_{opt}) = \frac{(I(f, \Sigma, \Omega))^{p'}}{p' m^{1/(p-1)}}.$$

The quantity $I(f, \Sigma, \Omega)$ can also be related to the dual problem (16) by introducing the 1-homogeneous function $\rho(z)$ as

$$\rho(z) = \inf \{t > 0 \; : \; j(z/t) \leq 1/p\};$$

so that we have

$$j(z) = \frac{1}{p} (\rho(z))^p.$$

The polar function associated to ρ is given by:

$$\rho^0(z) = \sup \{z : \xi \; : \; \rho(\xi) \leq 1\},$$

and we obtain

$$j^*(z) = \frac{1}{p'} (\rho^0(z))^{p'} \qquad \forall z \in \mathbb{R}^{N \times N}.$$

In this way we have the equality

$$I(f, \Sigma, \Omega) = \inf \left\{ \int \rho^0(\lambda) \; : \; \lambda \in \mathcal{M}(\mathbb{R}^N; \mathbb{R}^{N \times N}), \right.$$

$$\left. \mathrm{spt}\lambda \subset \overline{\Omega}, \; -\mathrm{div}\lambda = f \text{ in } \mathcal{D}'(\mathbb{R}^N \setminus \Sigma; \mathbb{R}^N) \right\}, \quad (19)$$

where the integral is intended in the sense of convex functionals on measures, and the following result holds (see [23]).

Proposition 3.2. *If μ is a solution of the mass optimization problem (15) then one has*

$$I(f, \Sigma, \Omega) = \min \left\{ \int \rho^0(\sigma) \, d\mu \; : \; \sigma \in L^1_\mu(\mathbb{R}^N; \mathbb{R}^{N \times N}), \right.$$

$$\left. -\mathrm{div}(\sigma\mu) = f \text{ in } \mathcal{D}'(\mathbb{R}^N \setminus \Sigma; \mathbb{R}^N) \right\} \quad (20)$$

and every optimal σ in (20) verifies

$$\rho^0(\sigma) = \frac{I(f, \Sigma, \Omega)}{m} \qquad \mu\text{-almost everywhere.} \quad (21)$$

Conversely, if λ is a solution of (19), then the nonnegative measure

$$\mu := \frac{m}{I(f, \mathcal{U}, \Omega)} \rho^0(\lambda)$$

is optimal for (15).

Remark 3.3. It is interesting to notice that by Proposition 3.1 the optimal mass distributions for problem (15) can be deduced from solutions of problem (19), hence they do not depend on the growth exponent p of the energy density j but only on the convex level set $\{z \in \mathbb{R}^{N \times N} \ : \ \rho(z) \leq 1\}$. Moreover, when μ is an optimal mass distribution, by (21) the associated stress density $j^*(\sigma) = \frac{1}{p'}\left(\rho^0(\sigma)\right)^{p'}$ is constant.

The optimal mass distribution problem (15) can be equivalently rephrased in terms of a PDE that we call Monge-Kantorovich equation. In order to deduce this equation from the mass optimization problem, it is convenient to introduce the class $\mathrm{Lip}_{1,\rho}(\Omega, \Sigma)$ as the closure, in $C(\overline{\Omega}; \mathbb{R}^N)$, of the set $\{u \in \mathcal{D}(\mathbb{R}^N; \mathbb{R}^N) \ : \ \rho(Du) \leq 1 \text{ on } \Omega, \ u = 0 \text{ on } \Sigma\}$. We notice that when $\rho(z) \geq |z|$ then every function in $\mathrm{Lip}_{1,\rho}(\Omega)$ is locally Lipschitz continuous on Ω; on the other hand, if $\rho(z) = |z^{sym}|$ it is known that this is no more true, due to the lack of Korn inequality for $p = +\infty$ (see for instance [94]).

We can now define the relaxed formulation of problem (17) as

$$\sup\left\{\langle f, u \rangle \ : \ u \in \mathrm{Lip}_{1,\rho}(\Omega, \Sigma)\right\} \tag{22}$$

and the finite dimensional linear space of all rigid displacements vanishing on Σ

$$\mathcal{R}_\Sigma = \left\{u(x) = Ax + b \ : \ b \in \mathbb{R}^N, \ A \in \mathbb{R}^{N \times N}_{skew}, \ u = 0 \text{ on } \Sigma\right\}.$$

Proposition 3.3. *The supremum in problem (17) is finite if and only if*

$$\langle f, u \rangle = 0 \qquad \forall u \in \mathcal{R}_\Sigma. \tag{23}$$

In this case, problem (22) admits a solution and

$$\sup(17) = \max(22).$$

Remark 3.4. In the scalar case, with $\rho(z) = |z|$, it is easy to see that the class $\mathcal{R}_\Sigma$ reduces to the function identically zero when $\Sigma \neq \emptyset$ and to the family of constant functions when $\Sigma = \emptyset$. Therefore, in the scalar case the quantity $I(f, \Sigma, \Omega)$ is always finite whenever $\Sigma \neq \emptyset$ or, if $\Sigma = \emptyset$, provided the source f has zero average.

In order to well define the optimality conditions for problem (15) and to deduce the Monge-Kantorovich PDE, we need to introduce the function space of displacements of finite energy related to a general measure μ. We refer to [27] and [28] for a complete presentation of the tools we shall use; in particular for the notion of tangent bundle of a measure μ, generalizing the classical one of k-dimensional manifolds S, which correspond in this framework to the measures of the form $\mu = \mathcal{H}^k \llcorner S$.

Given a measure μ and an open subset U of $\mathbb{R}^N$ we define the space of admissible stresses

$$X_\mu^{p'}(U; \mathbb{R}_{sym}^{N \times N}) = \left\{ \sigma \in L_\mu^{p'}(U; \mathbb{R}_{sym}^{N \times N}) \; : \; \mathrm{div}(\sigma\mu) \in \mathcal{M}(\mathbb{R}^N; \mathbb{R}^N) \right\}$$

and the tangent set of matrices

$$M_\mu(x) = \mu - ess\bigcup \left\{ \sigma(x) \; : \; \sigma \in X_\mu^{p'}(U; \mathbb{R}_{sym}^{N \times N}) \right\},$$

where $\mu - ess$ stands for the μ essential union. If $P_\mu(x)$ denotes the orthogonal projector on $M_\mu(x)$ with respect to the usual scalar product on matrices, for every function $u \in \mathcal{D}(U; \mathbb{R}^N)$ we may then define the tangential strain $e_\mu(u)$ as

$$e_\mu(u)(x) = P_\mu(x) \, Du(x).$$

It is possible to show that the linear operator

$$u \in \mathcal{D}(U; \mathbb{R}^N) \mapsto e_\mu(u) \in L_\mu^p(U; \mathbb{R}_{sym}^{N \times N})$$

is closable as an operator from $C(\overline{U}; \mathbb{R}^N)$ into $L_\mu^p(U; \mathbb{R}_{sym}^{N \times N})$, and we still denote by e_μ the closed operator from $C(\overline{U}; \mathbb{R}^N)$ into $L_\mu^p(U; \mathbb{R}_{sym}^{N \times N})$ which extends the tangential strain.

Now we can define the Banach space of all finite energy displacements $\mathcal{D}_{0,\mu}^{1,p}(U)$ as the domain of the operator e_μ endowed with the norm

$$\|u\|_{\mathcal{D}_{0,\mu}^{1,p}(U)} = \|u\|_{C(\overline{U})} + \|e_\mu(u)\|_{L_\mu^p(U)}.$$

It is then possible to obtain the relaxed form of the stored energy functional $J(\mu, u)$

$$\overline{J}(\mu, u) = \inf \left\{ \liminf_{h \to +\infty} J(\mu, u_h) \; : \; u_h \to u \text{ uniformly, } u_h \in \mathcal{D}(U; \mathbb{R}^N) \right\} =$$
$$= \begin{cases} \int_U j_\mu(x, e_\mu(u)) \, d\mu & \text{if } u \in \mathcal{D}_{0,\mu}^{1,p}(U) \\ +\infty & \text{otherwise.} \end{cases}$$

$$(24)$$

where

$$j_\mu(x, z) = \inf \left\{ j(z + \xi) \; : \; \xi \in \left(M_\mu(x)\right)^\perp \right\}.$$

This allows us to prove that for every nonnegative measure μ in $\overline{\Omega}$ and for every admissible displacement $u \in \mathrm{Lip}_{1,\rho}(\Omega, \Sigma)$ we have that $u \in \mathcal{D}_{0,\mu}^{1,p}(\mathbb{R}^N \setminus \Sigma)$ and $j_\mu(x, e_\mu(u)) \le 1/p$ μ-a.e. on $\mathbb{R}^N \setminus \Sigma$.

The scalar case is slightly simpler; indeed we define

$$X_\mu^{p'}(U; \mathbb{R}^N) = \left\{ \sigma \in L_\mu^{p'}(U; \mathbb{R}^N) \; : \; \mathrm{div}(\sigma\mu) \in \mathcal{M}(\mathbb{R}^N; \mathbb{R}^N) \right\}$$

and the tangent space $T_\mu(x)$ for μ-a.e. x as

$$T_\mu(x) = \mu - ess\bigcup \left\{ \sigma(x) \; : \; \sigma \in X_\mu^{p'}(U; \mathbb{R}^N) \right\}.$$

The orthogonal projector $P_\mu(x)$ on $T_\mu(x)$ and the tangential gradient $D_\mu u$ are defined similarly as above, as well as the relaxed energy density which becomes

$$j_\mu(x,z) = \inf\left\{ j(z+\xi) \ : \ \xi \in (T_\mu(x))^\perp \right\}.$$

Note that in the case $j(z) = |z|^2/2$ we obtain $j_\mu(x,z) = |P_\mu(x)z|^2/2$. Moreover, in this case $\rho(z) = |z|$ and we have

$$\mathrm{Lip}_{1,\rho}(\Omega; \Sigma) = \left\{ u \in W^{1,\infty}(\Omega) \ : \ u = 0 \text{ on } \Sigma, \ |Du| \le 1 \text{ a.e. on } \Omega \right\}.$$

We are now in a position to introduce the Monge-Kantorovich equation that can be written as:

$$\begin{cases} -\mathrm{div}(\sigma\mu) = f & \text{on } \mathbb{R}^N \setminus \Sigma \\ \sigma \in \partial j_\mu\big(x, e_\mu(u)\big) & \mu\text{-a.e. on } \mathbb{R}^N \\ u \in \mathrm{Lip}_{1,\rho}(\Omega, \Sigma) \\ j_\mu\big(x, e_\mu(u)\big) = 1/p \ \mu\text{-a.e. on } \mathbb{R}^N \\ \mu(\Sigma) = 0. \end{cases} \tag{25}$$

where $\partial j_\mu(x, \cdot)$ denotes the subdifferential of the convex function $j_\mu(x, \cdot)$. The link between the mass optimization problem (15) and the Monge-Kantorovich equation above has been investigated in [23]; we summarize here the result.

Theorem 3.2. *If μ solves the mass optimization problem (15) and u and σ are solutions of problems (22) and (20) respectively, then the triple $(u, m\sigma/I, I\mu/m)$ solves the Monge-Kantorovich equation (25) with $I = I(f, \Sigma, \Omega)$. Vice versa, if the triple (u, σ, μ) solves the Monge-Kantorovich equation (25), then u is a solution of problem (22) and the measure $m\mu/I$ is a solution of the mass optimization problem (15). Moreover, $I\sigma/m$ is a solution of the stress problem (16), and $(I/m)^{1/(p-1)}u$ is a solution of the relaxed displacement problem*

$$\min\left\{ \overline{J}(\mu, v) - \langle f, v \rangle \ : \ v \in \mathcal{D}^{1,p}_{0,\mu}(\mathbb{R}^N \setminus \Sigma) \right\}, \tag{26}$$

both related to the measure $m\mu/I$.

The scalar case is again simpler; indeed, by also taking $\rho(z) = |z|$ the Monge-Kantorovich equation (25) becomes

$$\begin{cases} -\mathrm{div}(\mu D_\mu u) = f & \text{on } \mathbb{R}^N \setminus \Sigma \\ u \in W^{1,\infty}(\Omega), \ u = 0 \text{ on } \Sigma, \ |Du| \le 1 \text{ a.e. on } \Omega \\ |D_\mu u| = 1 & \mu\text{-a.e. on } \mathbb{R}^N \\ \mu(\Sigma) = 0. \end{cases} \tag{27}$$

4 Optimal transportation problems

In this section we introduce the optimal mass transportation problem in his strong form (Monge problem) and in his weak or relaxed form (Kantorovich problem). In Section 5 we will enlight some full or partial equivalences with the mass optimization problems introduced in the previous section.

This equivalence will permit to give in the scalar case an explicit formula for the optimal mass and to deduce some regularity results as well as to allow some useful variational approximations of the mass optimization problem.

4.1 The optimal mass transportation problem: Monge and Kantorovich formulations

In order to have a general framework, in this section we will work in a locally compact metric space (X, d). This will make more clear various notions and will permit an unified treatment of different situations when studying the relationships with the mass optimization problems. The problem was originally formulated by Monge in 1781 (see [35]) in the Euclidean space. Using a modern terminology we will use measures where Monge was generically speaking of mass densities. This also permits a larger flexibility of the model.

Let f^+ and f^- be two probability measures on X; a *transport map* of f^+ on f^- is an element of the set

$$T(f^+, f^-) := \{\varphi : X \to X \text{ s.t. } \varphi \text{ is measurable and } \varphi_\sharp f^+ = f^-\} \qquad (28)$$

where $\varphi_\sharp f^+$ denote the push forward of f^+ through φ i.e. the measure on X defined as follows:

$$\varphi_\sharp f^+(B) := f^+(\varphi^{-1}(B)) \qquad \text{for all Borel subsets } B \text{ of } X.$$

A general (justified by physical or economical applications) cost of a transport map in $T(f^+, f^-)$ is given by

$$J(\varphi) = \int_X \psi\big(d(x, \varphi(x))\big) \, df^+(x). \qquad (29)$$

The quantity $\psi\big(d(x, \varphi(x))\big) \, df^+(x)$ represents the cost of transportation for mass unity. Then $\psi : \mathbb{R}^+ \to \mathbb{R}^+$ is a positive increasing function. In the original formulation Monge considered $\psi(t) = t$, while in economical applications it is reasonable to expect that ψ is concave (see for instance [106]); on the other hand, strictly convex functions ψ are used to study certain classes of differential equations (see for instance [12], [140], [154]).

Then a general formulation of the Monge problem is:

$$\min\left\{ \int_X \psi\big(d(x, \varphi(x))\big) \, df^+(x) \ : \ \varphi \in T(f^+, f^-)\right\}. \qquad (30)$$

Remark 4.1. It may happen that the class $T(f^+, f^-)$ is empty; this occurs for instance when $f^- = \frac{1}{2}(\delta_{x_0} + \delta_{x_1})$ and $f^+ = \delta_{x_2}$ in a space X which contains at least 3 points. Furthermore, even if $T(f^+, f^-) \neq \emptyset$, in general the minimum in (30) is not achieved as shown in the next example.

Example 4.1. Denote by I the interval $[0,1]$ and consider as X the Euclidean plane $\mathbb{R}^2$, $f^+ = \mathcal{H}^1 \llcorner (\{0\} \times I)$ and $f^- = \frac{1}{2}\mathcal{H}^1 \llcorner (\{1\} \times I) + \frac{1}{2}\mathcal{H}^1 \llcorner (\{-1\} \times I)$. In this case it is easy to see that the infimum in problem (30) is 1 but for each transport map φ we have $|x - \varphi(x)| \geq 1$ for f^+ a.e. x and there is a set of positive f^+-measure for which $|x - \varphi(x)| > 1$.

The behaviour of minimizing sequences of transport maps in the previous example shows also that the class of transport maps is not closed with respect to a topology sufficiently strong to deal with the nonlinearity of functional (29). Then it is natural to think to a possible relaxed formulation of the problem. The relaxed formulation we are introducing is due to Kantorovich and it consists in embedding the class of transport maps in the larger class of the so called *transport plans* which will be introduced below.

A transport plan of f^+ to f^- is a probability measure $\gamma \in \mathcal{M}(X \times X)$ which belongs to the set:

$$P(f^+, f^-) = \{\gamma \in \text{Prob}(X \times X) \ : \ \pi^1_\sharp \gamma = f^+, \ \pi^2_\sharp \gamma = f^-\}, \qquad (31)$$

where for all $(x,y) \in X \times X$, $\pi^1(x,y) = x$ and $\pi^2(x,y) = y$ are the projections on the first and second factors of $X \times X$. It is easy to check that (denoting by id the identity map) $(id \otimes \varphi)_\sharp f^+ \in P(f^+, f^-)$ if and only if $\varphi \in T(f^+, f^-)$. The reader familiar with the Young measures will recognize in the transport plan associated to a transport map φ the measure associated to the graph of φ. As the embedding of $T(f^+, f^-)$ in $P(f^+, f^-)$ is given by $\varphi \mapsto (id \otimes \varphi)_\sharp f^+$ the natural extension of the cost functional is given by

$$J(\gamma) = \int_{X \times X} \psi(d(x,y) d\gamma(x,y), \qquad (32)$$

so that Monge problem (30) relaxes to the Kantorovich problem:

$$\min \left\{ \int_{X \times X} \psi\big(d(x,y)\big) \, d\gamma(x,y) \ : \ \gamma \in P(f^+, f^-) \right\}. \qquad (33)$$

Remark 4.2. The class $P(f^+, f^-)$ is never empty as it always contains $f^+ \otimes f^-$. Moreover $P(f^+, f^-)$ is closed with respect to the tight convergence on the space of finite measures. This implies that problem (33) admits a minimizer whenever the function ψ is lower semicontinuous.

Example 4.2. Let f^+ and f^- be the measures of Example 4.1, then it is easy to see that the only optimal transport plan is given by the probability measure on $\mathbb{R}^4$

$$\gamma = f^+ \otimes f^-$$

which turns out to verify for every test function $\theta(x, y, z, w)$ on $\mathbb{R}^4$

$$\langle \gamma, \theta \rangle = \frac{1}{2} \int_0^1 \int_0^1 [\theta(0, y, 1, w) + \theta(0, y, -1, w)] \, dy \, dw \ .$$

Therefore again we conclude that the minimum cannot be achieved on the set of tranport maps.

We go now back to consider a bounded and connected open subset Ω of $\mathbb{R}^N$ and a closed subset Σ of $\overline{\Omega}$. We want to define a semi-distance on $\overline{\Omega}$ that only allows paths in $\overline{\Omega}$ and that does not count the lenght of paths along Σ. We start with the case when Σ is empty: in this case we simply define d_Ω as the geodesic distance in $\overline{\Omega}$, that is

$$d_\Omega(x,y) = \inf\left\{ \int_0^1 |\xi'(t)|\, dt \ : \ \xi(0) = x,\ \xi(1) = y,\ \xi(t) \in \overline{\Omega}\right\}.$$

If $\Sigma \neq \emptyset$ we set

$$d_{\Omega,\Sigma}(x,y) = \min\left\{d_\Omega(x,y), d_\Omega(x,\Sigma) + d_\Omega(y,\Sigma)\right\} \tag{34}$$

where we denoted, as usual, by $d_\Omega(x,\Sigma)$ the d_Ω-distance of the point x from the closed set Σ.

With the notation above the Kantorovich problem, for the choice $\psi(t) = t$, can be written as:

$$\min\left\{ \int_{\Omega\times\Omega} d_{\Omega,\Sigma}(x,y)\, d\gamma(x,y) \ : \ \gamma \in P(f^+, f^-)\right\}. \tag{35}$$

By denoting by $Lip_1(\Omega, d_{\Omega,\Sigma})$ the class of functions defined in Ω that are 1-Lipschitz continuous with respect to the distance $d_{\Omega,\Sigma}$ one may obtain a dual formulation of problem (35) in the form (see [23]):

$$I(f, \Sigma, \Omega) := \max\left\{\langle f, u\rangle \ : \ u \in Lip_1(\Omega, d_{\Omega,\Sigma}),\ u = 0 \text{ on } \Sigma\right\}. \tag{36}$$

4.2 The PDE formulation of the mass transportation problem

A system of partial differential equations can be associated to the Monge-Kantorovich problem (see for instance [23], [99], [100]). In a simplified version (i.e. Ω convex and bounded, Σ empty and f^+ and f^- regular functions with separated supports) this system was first used to prove the existence of an optimal transport by Evans and Gangbo in [100] and it can be written as follows:

$$\begin{cases} -\operatorname{div}\big(a(x)Du(x)\big) = f \text{ in } \Omega \\ u \in Lip_1(\Omega) \\ |Du| = 1 \qquad\qquad a(x)\text{-a.e.} \end{cases} \tag{37}$$

The geometrical informations about the directions in which the mass has to be transported in order to minimize the cost are contained in the gradient of u, while the density of the *transport rays* is contained in the coefficient $a(x)$.

The interest of this system in the mass optimization theory is due to the fact that it can be used to characterize the solutions of problem (15). We will show more in the next section.

5 Relationships between optimal mass and optimal transportation

We begin to observe that as j is 2-homogeneous it can be written as $j(z) = \frac{1}{2}(\rho(z))^2$ where ρ is positive, convex and 1-homogeneous. Then, as already done in (22), we consider the quantity

$$I(f, \Sigma, \Omega) = \max\left\{ \langle f, u \rangle \ : \ u \in Lip_{1,\rho}(\Omega, \Sigma) \right\}. \tag{38}$$

Remark 5.1. In the scalar case (with Ω convex and $\Sigma = \emptyset$, otherwise the same modifications seen above apply, see [23] for a general discussion) the interpretation of (38) in terms of transportation problem is related to the distance defined by

$$d_\rho(x, y) = \inf\left\{ \int_0^1 \rho^0(\xi'(t))\, dt \ : \ \xi(0) = x,\ \xi(1) = y \right\},$$

where $\rho^0(z) := \sup\{w \cdot z \ : \ \rho(w) \le 1\}$.

Using the duality between continuous functions and measures one obtains the identity:

$$I(f, \Sigma, \Omega) = \inf\left\{ \int \rho^0(\lambda) \ : \ \lambda \in \mathcal{M}(\overline{\Omega}),\ -\mathrm{div}\lambda = f \ \text{in} \ \mathbb{R}^N \setminus \Sigma \right\}. \tag{39}$$

Here the integral $\int \rho^0(\lambda)$ is intended in the sense of convex functions of a measure; more precisely, if θ is a convex function and λ is a vector measure, we set

$$\int \theta(\lambda) := \int \theta(\lambda^a)\, dx + \int \theta^\infty\Big(\frac{d\lambda^s}{d|\lambda^s|}\Big) d|\lambda^s|$$

where θ^∞ is the recession function of θ and $\lambda = \lambda^a\, dx + \lambda^s$ is the decomposition of λ into its absolutely continuous part λ^a and singular part λ^s. Finally we remark that if θ is 1-homogeneous (as for instance in the case of $\theta = \rho^0$) then θ coincides with θ^∞.

Propositions 3.1 and 3.2 establish a relation between problem (38) and the mass optimization problem in both the scalar and vectorial case. Theorem 3.2 shows the relation between the Monge-Kantorovich equations (25) ((27) in the scalar case) and the solutions of the mass optimization problem in both the scalar and the vectorial case (see [23] for the detailed proofs).

The equivalence between (38) and the transportation problem in the scalar case is based on duality again, and it goes back to Kantorovich. We refer to the lectures by Evans [99] and to the work by Evans and Gangbo [100] for the proof of this equivalence, at least in the case $\Omega = \mathbb{R}^N$, $\Sigma = \emptyset$, $\rho(z) = |z|$, and f satisfying suitable regularity conditions. The general case has been considered by Bouchitté and Buttazzo in [23]. In the vectorial case, even if the equivalence between problem (38), the mass optimization problem, and the

Monge-Kantorovich equations (25) still holds, it seems that no transportation problem can be identified. In other words, in the vectorial case, the mass optimization problem (15) and the Monge-Kantorovich equations (25) do not seem to be related to the mass transportation with respect to a distance.

For the remaining part of this section we restrict our attention to the scalar case, to a convex open set Ω and to $j(x) = \frac{1}{2}|z|^2$. More general situatios are considered in [23] (see also [64]).

Given an optimal transport plan γ it is possible to deduce an optimal density μ for the mass optimization problem (15) through the formula

$$\mu(B) = c \int_{\Omega \times \Omega} \mathcal{H}^1_{xy}(B) \, d\gamma(x, y) \tag{40}$$

where c is a suitable constant and $\mathcal{H}^1_{xy}$ denotes the 1-dimensional measure on the segment joining x and y.

Vice versa, it is possible to show (see [7]) that given a solution μ of the mass optimization problem (hence a measure λ which solves (19)) there exists an optimal transport plan γ such that μ is associated to γ through formula (40) above.

Remark 5.2. A formula like (40) still holds in the case of nonconvex domains Ω, and also in presence of a Dirichlet region Σ and with different costs j. The difference is that instead of considering the segment joining x to y one has to consider a geodesic g_{xy} for the distance $d_{\Omega, \Sigma}$ introduced in the previous section.

Formula (40) often permits to deduce several properties of the optimal masses and sometimes to identify them directly from the data f^+ and f^- (see [23] for some explicit examples). For instance, if we define the geodesic envelope $G(f^+, f^-)$ of the support of f as the union of all geodesic lines joining a point in the support of f^+ to a point in the support of f^- we have the following result.

Theorem 5.1. *Let μ be a solution for the mass optimization problem (15). Then the support of μ is a subset of $G(f^+, f^-)$. In particular, if Ω is convex and $\Sigma = \emptyset$, then*

$$\mathrm{spt}\,\mu \subset \mathrm{co}\big(\mathrm{spt}\,f^+ \cup \mathrm{spt}\,f^-\big), \tag{41}$$

where co denotes the convex envelope operation.

The following regularity results have been obtained using formula (40) and a property of the support of optimal plans called ciclical monotonicity (see [96], [97]).

Theorem 5.2. *Let μ be a solution for the mass optimization problem 15. Then $\dim(\mu) \geq \max\big\{\dim f^+, \dim f^-, 1\big\}$. In particular, if f^+ (or f^-) is absolutely continuous so is μ. Moreover for every $p \in [1, +\infty]$*

$$f^+, f^- \in L^p \;\Rightarrow\; \mu \in L^p \,.$$

Remark 5.3. The following facts merit to be emphasized:

- It was proved indipendently in [7] and [102] that the optimal density μ is unique whenever $f^+, f^- \in L^1$. We stress that a similar result is not known in the vectorial case.
- The first summability result for μ was obtained in [96], where for $1 < p < +\infty$ only the $L^{p-\varepsilon}$ regularity was proved. The summability improvement has been recently obtained in [97] by means of elliptic PDE estimates.
- If only one of the two measures f^+ and f^- is in L^p then it can be proved that $\mu \in L^q$ for slightly more complicated q.
- Nothing is known about the regularity of μ in the vectorial case.

6 Further results and open problems

6.1 A vectorial example

Here we present an example of application of the theory developed in the previos sections to a problem of optimal structures in elasticity. We consider the linear isotropic stored energy

$$j(z) = \frac{\alpha}{2} |tr(z^{sym})|^2 + \beta |z^{sym}|^2$$

where α and β are the Lamé constants in dimension two.

The problem is the following: distribute in $\mathbb{R}^2$ a given amount of mass in order to minimize the elastic compliance related to the force field $f = \delta_A \tau_1 + \delta_B \tau_2 + \delta_C \tau_3$ which is described in the figure below

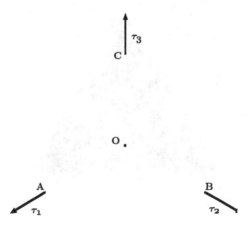

Fig. 3. The force field f.

A natural guess is that one of the two structures in the figure below is optimal.

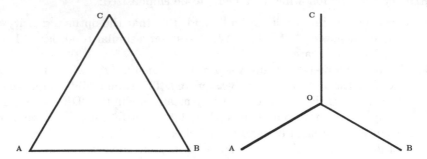

Fig. 4. Two admissible structures.

Indeed it is possible to show that the elastic compliance associated to the two structures is the same and that it is the lowest possible among all 1-dimensional structures. However, none of the two choices is optimal, since a multiple of the optimal mass distribution should satisfy the Monge-Kantorovich equation and this does not occur for the two structures above (see [23] for a proof). A numerical computation of a two dimensional optimal structure can be found in [109], and the optimal distribution is represented in the figure below.

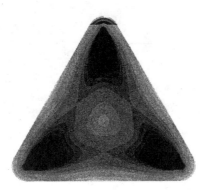

Fig. 5. The optimal mass distribution.

Remark 6.1. The example above shows that Theorem 5.1 does not hold in the vectorial case. A weaker formulation, which could be expected to hold in the vectorial case, is:

$$\mathrm{spt} f \text{ compact} \quad \Rightarrow \quad \mathrm{spt} \mu \text{ compact.}$$

We do not know to prove (or disprove) this property.

6.2 A p–Laplacian approximation

We now discuss some approximation of the mass optimization problem which is naturaly suggested by Remark 3.3. Indeed, as the optimal mass does not depend on the growth exponent of j, which we recall can be written as $j(z) = \frac{1}{p}\rho(z)^p$, it is natural to study the asymptotics for $p \to \infty$ of

$$F_p(u) = \begin{cases} \frac{1}{p} \int_\Omega \rho(e(u))^p \, dx - \langle f, u \rangle & \text{if } u \in W^{1,p}(\Omega, \mathbb{R}^N) \text{ and } u = 0 \text{ on } \Sigma, \\ +\infty & \text{otherwise.} \end{cases}$$

$$(42)$$

What we espect is that problem

$$\min_u F_p(u) \tag{43}$$

converges, when $p \to +\infty$, to

$$\min \big\{ -\langle f, u \rangle \ : \ \rho(e(u)) \le 1 \text{ a.e., } u = 0 \text{ on } \Sigma \big\}, \tag{44}$$

whose relationships with the mass optimization problem have been explained in the previous sections of these notes. As the dual problem of (44) is also involved in the mass optimization problem, it is natural to look also at the asymptotic behaviour of the dual problem of (43) which is:

$$\min \Big\{ \frac{1}{p'} \int_\Omega (\rho^0(\sigma))^{p'} \, dx \ : \ \sigma \in L^{p'}(\Omega, \mathbb{R}^{N \times N}), \ -\mathrm{div}\sigma = f \text{ in } \Omega \setminus \Sigma \Big\}. \tag{45}$$

Notice that the duality relationships says that if u_p is a minimizer of F_p then any minimizer of (45) can be written in the form $\sigma_p = \xi_p \rho(e(u_p))^{p-2}$ where ξ_p is an element of $\partial(\rho(e(u_p)))$. The convergence result proved in [24] is:

Theorem 6.1. *Assume that the minimum in (44) is finite and let $\{(u_p, \sigma_p)\}$ be a sequence of minimizers for (43) and (45). Set:*

$$\mu_p := \rho(e(u_p))^{p-2}, \quad \sigma_p = \xi_p \, \mu_p \ .$$

Then there exist $u \in Lip_{1,\rho}(\Omega, \Sigma)$, $\mu \in \mathcal{M}(\overline{\Omega})$ and a suitable rigid displacement $r_p \in \mathcal{R}_\Sigma$, such that, up to subsequences:

(i) $u_p - r_p \to u$ uniformly;

(ii) $\mu_p \rightharpoonup \mu$ *in* $\mathcal{M}(\overline{\Omega})$;
(iii) $\xi_p \mu_p \rightharpoonup \xi \mu$ *in* $\mathcal{M}(\overline{\Omega}, \mathbb{R}^N)$;
(iv) u *is a minimizer of (44) and* (u, ξ, μ) *is a solution of the Monge-Kantorovich equation.*

Moreover, if the set $\{\rho^0 \leq 1\}$ *is strictly convex, then the convergence* $\xi_p \rightarrow \xi$ *holds in a stronger sense.*

Remark 6.2. In the scalar case of conductivity, Theorem 6.1 can be reformulated as follows: assume that $\Sigma \neq \emptyset$ (or that $\int df = 0$) and let u_p be the unique solution (up to an additive constant) of:

$$-\Delta_p u_p = f \quad \text{on } \Omega \setminus \Sigma, \quad u_p = 0 \quad \text{on } \Sigma, \quad \frac{\partial u_p}{\partial \nu} = 0 \quad \text{on } \partial\Omega \setminus \Sigma.$$

Then up to subsequences, we have:

$$u_p \rightarrow u \quad \text{uniformly,}$$
$$|\nabla u_p|^{p-2} \rightharpoonup \mu \quad \text{in } \mathcal{M}(\overline{\Omega}),$$
$$|\nabla u_p|^{p-2}\nabla u_p \rightharpoonup \mu \nabla_\mu u \quad \text{in } \mathcal{M}(\overline{\Omega}, \mathbb{R}^N),$$

where (u, μ) solves the scalar Monge-Kantorovich equation.

Remark 6.3. In the scalar case and for f regular enough the approximation above has have been used by Evans and Gangbo in [100] to prove one of the first existence results for the Monge problem.

Remark 6.4. The main difference between the scalar and the vectorial case is that in the vectorial case one has to assume that the limit problem admits a minimizer to obtain the approximation result (6.1) while in the scalar case this is a consequence of the approximation. This is due to the fact that the Korn's inequality (needed in the vectorial case) does not hold in $W^{1,\infty}$ and then it is not stable in $W^{1,p}$ when $p \rightarrow \infty$. In the scalar case the main role is played by the Poincaré inequality which is more stable.

6.3 Optimization of Dirichlet regions

If we consider the cost in (33), (35)

$$J(\Sigma) = \min\left\{ \int_{\overline{\Omega} \times \overline{\Omega}} \psi(d_{\Omega,\Sigma}(x,y))\, d\gamma(x,y) \; : \; \gamma \in P(f^+, f^-) \right\} \tag{46}$$

as a function of Σ only, once f^+, f^-, Ω, ψ are fixed, a natural question which arises is to optimize the functional J in a suitable class of admissible Dirichlet regions Σ. To this kind of problems belong several questions, known in the literature under different names. We list here below some of them.

Location problems. In this setting $f^- \equiv 0$ and $f^+ \in L^1(\Omega)$, so that the functional $J(\Sigma)$ in (46) simply becomes

$$J(\Sigma) = \int_\Omega \psi\big(\mathrm{dist}_\Omega(x, \Sigma)\big) f^+(x)\, dx. \qquad (47)$$

The admissible Σ are the subsets of $\overline{\Omega}$ made of n points; hence the optimal location problem reads as

$$\min\left\{ \int_\Omega \psi\big(\mathrm{dist}_\Omega(x, \Sigma)\big) f^+(x)\, dx \ : \ \Sigma \subset \overline{\Omega}, \ \#\Sigma = n \right\} \qquad (48)$$

where $\mathrm{dist}_\Omega(x, \Sigma)$ is the distance of x to Σ measured in the geodesic metric of Ω. Problem (48) describes the question of locating in a given region Ω a given number n of points of distribution of a certain product, in order to minimize the average distance (in the case $\psi(d) = d$) that customers, whose density $f^+(x)$ in Ω is known, have to cover to reach the closest point of distribution.

The existence of an optimal Σ_n, for a fixed n, is straightforward. It is also clear that the minimum value I_n of problem (48) tends to zero as $n \to +\infty$, and a simple argument allows us to obtain that

$$I_n = O\big(\psi(n^{-1/N})\big).$$

Hence it is interesting to study the asymptotic behaviour (as $n \to +\infty$) of the rescaled functionals

$$J_n(\Sigma) = \begin{cases} J(\Sigma)/I_n & \text{if } \#\Sigma = n \\ +\infty & \text{otherwise} \end{cases}$$

in terms of the Γ-convergence, with respect to the convergence defined on the space of measures by

$$\Sigma_n \to \lambda \quad \Longleftrightarrow \quad \frac{1}{n}\mathcal{H}^0 \llcorner \Sigma_n \to \lambda \quad \text{weakly}^*.$$

A reasonable conjecture, in the case Ω convex and $\psi(d) = d$, also supported by the results of [132], is that the Γ-limit functional is written as

$$J_\infty(\lambda) = C \int f^+(x)\lambda^{-1/N}$$

intended in the sense of convex functionals over the measures, where C is a suitable constant. In particular, this would imply that the best location of n points in Ω has to be asymptotically proportional to $\big(f^+(x)\big)^{N/(N+1)}$, being $f^+(x)$ the density of population in Ω.

It would also be interesting to compute how good, with respect to the optimal choice Σ_n, is a random choice of n points in Ω.

Irrigation problems. Take again $f^- \equiv 0$ and $f^+ \in L^1(\Omega)$, which gives $J(\Sigma)$ as in (47). The admissible Σ are now the closed connected subsets of $\overline{\Omega}$ whose 1-dimensional Hausdorff measure $\mathcal{H}^1$ does not exceed a given number L. The optimization problem then reads as

$$\min \left\{ J(\Sigma) \; : \; \Sigma \subset \overline{\Omega}, \; \Sigma \text{ closed connected, } \mathcal{H}^1(\Sigma) \le L \right\} \qquad (49)$$

In other words we want to irrigate the region Ω by a system of water channels Σ and the cost of irrigating a point $x \in \Omega$, where a mass $f^+(x)dx$ is present, is assumed to be proportional to $\psi\big(\text{dist}_\Omega(x, \Sigma)\big)$.

The existence of an optimal Σ_L easily follows from the Golab theorem (see [64]) and several qualitative properties of Σ_L can be shown. We refer to [64] for a more complete analysis and open problems about the structure of Σ_L. Here we want simply to stress that an asymptotic study (as $L \to +\infty$), similar to the one presented above for location problems, seems interesting; however, if I_L denotes the minimum value of problem (49), it is still possible to obtain, at least when $\psi(d) = d$, the asymptotic estimate (see [64])

$$I_L = O(L^{1/(1-N)}).$$

This makes reasonable the conjecture that the Γ-limit functional (as $L \to +\infty$), with respect to a convergence on Σ_L similar to the one above, takes the form

$$J_\infty(\lambda) = K \int f^+(x) \lambda^{1/(1-N)}$$

for a suitable constant K. Again, this would imply that the density of the optimal irrigation channels is asymptotically (as $L \to +\infty$) proportional to $\big(f^+(x)\big)^{(N-1)/N}$.

Problems in urban planning. Here f^+ represents the density of the working population in an urban area Ω, and f^- the density of the working places, that are assumed to be known. Notice that in several problems in urban planning f^+ and f^- are the main unknowns and the optimal location of them in a given region Ω is obtained through the minimization of a cost functional which takes into account a penalization for concentrated f^+, a penalization for sparse f^-, and the cost of transporting f^+ onto f^-.

The closed connected subset Σ of $\overline{\Omega}$ represents the urban transportation network that has to be designed in order to minimize the cost $J(\Sigma)$ in (46). The problem then becomes

$$\min \left\{ J(\Sigma) \; : \; \Sigma \subset \overline{\Omega}, \; \Sigma \text{ closed connected, } \mathcal{H}^1(\Sigma) \le L \right\}. \qquad (50)$$

In this model the population is transported on Σ for free, but it is possible to consider similar models where customers pay a ticket which depends in some way on the length of the part of Σ they used.

6.4 Optimal transporting distances

A different kind of optimization problem related to mass transportation consists in considering the marginal measures f^+ and f^- as given and to searching a distance d which optimizes the transportation cost among all distances belonging to some admissible class. More precisely, we consider a domain Ω of $\mathbb{R}^N$, that for simplicity we take convex, and for every Borel coefficient $a(x)$ the corresponding Riemannian distance

$$d_a(x,y) = \inf \left\{ \int_0^1 a(\gamma(t))|\gamma'(t)|\, dt \; : \; \gamma(0) = x, \; \gamma(1) = y \right\}.$$

A small coefficient $a(x)$ in a region ω makes the transportation in ω easy, while on the contrary, large coefficients make the transportation more costly.

The admissible class we shall take into consideration is the class of distances of the form d_a where the coefficient a varies into

$$\mathcal{A} = \left\{ \alpha \leq a(x) \leq \beta, \; \int_\Omega a(x)\, dx \leq m \right\}$$

being α, β, m given constants, satisfying the compatibility condition

$$\alpha|\Omega| \leq m \leq \beta|\Omega|.$$

For every $a \in \mathcal{A}$ we consider the mass transportation cost

$$F(a) = \inf \left\{ \int_{\overline{\Omega} \times \overline{\Omega}} \psi(d_a(x,y))\, d\gamma(x,y) \; : \; \pi_\#^1 \gamma = f^+, \; \pi_\#^2 \gamma = f^- \right\} \qquad (51)$$

where $\psi : [0, +\infty[\to [0, +\infty[$ is a continuous and nondecreasing function.

The minimization problem for F on $\mathcal{A}$, which correspond to find the most favourable coefficient for the transportation of f^+ onto f^-, is somehow trivial. In fact, it is easy to show that

$$\inf \left\{ F(a) \; : \; a \in \mathcal{A} \right\} = F(\alpha).$$

On the contrary, the problem of maximizing the functional F on $\mathcal{A}$ seems interesting. In other words, we want to prevent as much as possible the transportation of f^+ onto f^-, making its cost as big as possible, playing on the Riemannian coefficient $a(x)$. Notice that the integral constraint $\int_\Omega a(x)\, dx \leq m$ forces us to use carefully the quantity m (a kind of total prevention power) at our disposal: we do not have to waste high coefficients in regions which are not essential for the mass transportation

The first attempt to find a solution of the maximization problem

$$\max \left\{ F(a) \; : \; a \in \mathcal{A} \right\} \qquad (52)$$

consists in trying to use the direct methods of the calculus of variations: this requires to introduce a topology on the class of admissible choices strong enough to have the lower semicontinuity of the cost functional and weak enough to have the compactness of minimizing sequences. In our case the best choice would be to relate the convergence of a sequence of coefficients (a_n) to the convergence of geodesic paths corresponding to the distances d_{a_n}; more precisely, since the geodesic paths could be nonunique, the definition has to be given in terms of Γ-convergence for the length functionals L_n with respect to d_{a_n}. Here we mean that for every distance d we may consider the length functional L_d defined on curves $\gamma : [0,1] \to \Omega$ by

$$L_d(\gamma) = \sup \left\{ \sum_{i=1}^{k} d\big(\gamma(t_i), \gamma(t_{i+1})\big) \; : \; t_1 < \cdots < t_k \right\}. \tag{53}$$

Thus we define the convergence τ on the class of geodesic distances (that is the ones whose value between two points coincides with the infimum of the lengths of all paths joining the two points) by setting

$$d_n \to d \text{ in } \tau \quad \Longleftrightarrow \quad L_{d_n} \to L_d \text{ in the } \Gamma\text{-convergence}$$

where we considered the family of curves γ endowed with the uniform convergence.

It is not difficult to show that with this choice of convergence the functional $F(a)$ turns out to be continuous on $\mathcal{A}$; moreover (see [58]) we have that a sequence d_n converges in τ to d if and only if d_n converges to d uniformly as a sequence of functions in $\Omega \times \Omega$ (see also [65], where the uniform convergence $d_n \to d$ is related to the Γ-convergence of the onedimensional Hausdorff measures $\mathcal{H}^1_{d_n} \to \mathcal{H}^1_d$). Therefore, due to the upper bound β, all distances d_a with $a \in \mathcal{A}$ are equi-Lipschitz continuous, so that Ascoli-Arzelà theorem allows to conclude that the class $\mathcal{A}$ is pre-compact for the τ convergence.

Unfortunately, we cannot conclude by the usual argument of the direct methods of the calculus of variations because the class $\mathcal{A}$ is not closed under τ. This is known since a long time (see [1]) through the simple example of a chessboard structure in $\mathbb{R}^2$ of side ε, with $a_\varepsilon(x) = \alpha$ on white cells and $a_\varepsilon(x) = \beta$ on black cells, whose associated distances d_{a_ε} converge, in the sense above, to a distance d which is not of Riemannian type. The explicit computation of this limit d can be easily made when the ratio β/α is large enough, and in this case we find that d is a Finsler distance generated by a Finsler metric $\phi(z)$ (a convex and 1-homogeneous function) whose unit ball is a regular octagone.

On the other hand, it is known (see [153]) that the class of geodesic distances coincides with the class of Finsler distances, through the derivation formula

$$\phi(x,z) = \limsup_{t \to 0} \frac{d(x+tz, x)}{t} \, .$$

More recently, in [29] it has been actually proved a density result for the distances of the form d_a, with $\alpha \le a \le \beta$, in the class of all Finsler distances d generated by metrics $\phi(x,z)$, which leads us to expect that the relaxed maximization problem associated to (52) is defined on the τ closure $\overline{\mathcal{A}}$ of $\mathcal{A}$, which is a subclass of Finsler distances on Ω:

$$\max\left\{\mathcal{F}(d) \ : \ d \in \overline{\mathcal{A}}\right\} \tag{54}$$

being $\mathcal{F}(d)$ defined as in (51) with d which replaces d_a. However, we are able to prove that the original maximization problem (52) actually admits an optimal unrelaxed solution, thanks to the following lemma (see [57]).

Lemma 6.1. *Let $\phi(x,z)$ be a Finsler metric which comes out as a limit of a sequence (a_n) of coefficients in the original admissible class $\mathcal{A}$. Then, if we define the largest eigenvalue $\Lambda_\phi(x)$ of $\phi(x,\cdot)$ as*

$$\Lambda_\phi(x) = \sup\left\{\phi(x,z) \ : \ |z| \leq 1\right\},$$

we have that

$$\int_\Omega \Lambda_\phi(x)\,dx \leq m.$$

The proof of the existence of an optimal unrelaxed solution for the original maximization problem (52) follows now easily.

Theorem 6.2. *The original maximization problem (52) admits an optimal solution.*

Proof. Take a maximizing sequence (a_n); according to what seen above we may find a subsequence (still denoted by the same indices) which τ converges to a Finsler distance d generated by a Finsler metric $\phi(x,z)$. We have then

$$\mathcal{F}(d) = \lim_{n\to+\infty} F(a_n).$$

By the definition of the largest eigenvalue $\Lambda_\phi(x)$ we have

$$\phi(x,z) \leq \Lambda_\phi(x)|z|.$$

Using now the fact that the cost functional is monotone increasing with respect to the distance functions, we may conclude that the coefficient $a(x) = \Lambda_\phi(x)|z|$, which is admissible for our problem thanks to Lemma 6.1, maximizes the functional F on the class $\mathcal{A}$.

Acknowledgements. These notes have been written as a development of a course given by G. B. at the CIME Summer School "Optimal Transportation and Applications" held in Martina Franca on September 2001. We wish to thank the CIME for the excellent organization and for the friendly and constructive atmosphere during the School. This work is also part of the European Research Training Network "Homogenization and Multiple Scales" (HMS2000) under contract HPRN-2000-00109.

References

1. E. ACERBI, G. BUTTAZZO: *On the limits of periodic Riemannian metrics.* J. Analyse Math., **43** (1984), 183–201.
2. E. ACERBI, G. BUTTAZZO: *Reinforcement problems in the calculus of variations.* Ann. Inst. H. Poincaré Anal. Non Linéaire, **3** (4) (1986), 273–284.
3. G. ALLAIRE: *Shape optimization by the homogeneization method.* Applied Mathematical Sciences **146**, Springer-Verlag, New York (2002).
4. G. ALLAIRE, E. BONNETIER, G. FRANCFORT, F.JOUVE: *Shape optimization by the homogenization method.* Numer. Math. **76** (1997), 27–68.
5. G. ALLAIRE, R. V. KOHN: *Optimal design for minimum weight and compliance in plane stress using extremal microstructures.* Europ. J. Mech. A/Solids, **12** (6) (1993), 839–878.
6. L. AMBROSIO: *Introduction to geometric measure theory and minimal surfaces* (Italian), Scuola Normale Superiore, Pisa (1997).
7. L. AMBROSIO: *Lecture notes on optimal transport problems.* Preprint n. 32, Scuola Normale Superiore, Pisa (2000).
8. L. AMBROSIO, G. BUTTAZZO: *An optimal design problem with perimeter penalization.* Calc. Var., **1** (1993), 55–69.
9. L. AMBROSIO, N. FUSCO, D. PALLARA: *Functions of bounded variation and free discontinuity problems.* Oxford Mathematical Monographs, Clarendon Press, Oxford (2000).
10. L. AMBROSIO, A. PRATELLI: *Existence and stability results in the L^1 theory of optimal transportation.* Preprint Scuola Normale Superiore, Pisa (2002).
11. L. AMBROSIO, P. TILLI: *Selected Topics on Analysis in Metric Spaces.* Scuola Normale Superiore, Pisa, (2000).
12. M. S. ASHBAUGH: *Open problems on eigenvalues of the Laplacian.* Analytic and geometric inequalities and applications, 13–28, Math. Appl., 478, Kluwer Acad. Publ., Dordrecht (1999).
13. H. ATTOUCH: *Variational Convergence for Functions and Operators.* Pitman, Boston (1984).
14. H. ATTOUCH, G. BUTTAZZO: *Homogenization of reinforced periodic one-codimensional structures.* Ann. Scuola Norm. Sup. Pisa Cl. Sci., **14** (1987), 465–484.
15. D. AZE, G. BUTTAZZO: *Some remarks on the optimal design of periodically reinforced structures.* RAIRO Modél. Math. Anal. Numér., **23** (1989), 53–61.
16. E. BAROZZI, E. H. A. GONZALEZ: *Least area problems with a volume constraint.* In "Variational methods for equilibrium problems of fluids", Astérisque **118**, (1984), 33–53.
17. M. BELLIEUD, G. BOUCHITTE: *Homogenization of elliptic problems in a fiber reinforced structure. Nonlocal effects.* Ann. Scuola Norm. Sup. Pisa Cl. Sci., **26** (1998), 407–436.
18. M. BELLONI, G. BUTTAZZO, L. FREDDI: *Completion by Gamma-convergence for optimal control problems.* Ann. Fac. Sci. Toulouse Math., **2** (1993), 149–162.
19. M. BELLONI, B. KAWOHL: *A paper of Legendre revisited.* Forum Math., **9** (1997), 655–667.
20. M. BELLONI, A. WAGNER: *Newton's problem of minimal resistance in the class of bodies with prescribed volume.* Preprint Università di Parma, Parma (2001).

21. J. D. BENAMOU, Y. BRENIER: *Mixed L^2-Wasserstein optimal mapping between prescribed density functions.* J. Optim. Theory Appl., **111** (2001), 255–271.

22. M. BENDSØE: *Optimal shape design as a material distribution problem.* Struct. Optim., **1** (1989), 193–202.

23. G. BOUCHITTE, G. BUTTAZZO: *Characterization of optimal shapes and masses through Monge-Kantorovich equation.* J. Eur. Math. Soc., **3** (2001), 139–168.

24. G. BOUCHITTE, G. BUTTAZZO, L. DE PASCALE: *A p-Laplacian approximation for some mass optimization problems.* J. Optim. Theory Appl., (to appear).

25. G. BOUCHITTE, G. BUTTAZZO, I. FRAGALÀ: *Convergence of Sobolev spaces on varying manifolds.* J. Geom. Anal., **11** (2001), 399–422.

26. G. BOUCHITTE, G. BUTTAZZO, I. FRAGALÀ: *Bounds for the effective coefficients of homogenized low dimensional structures.* J. Math. Pures Appl, (to appear).

27. G. BOUCHITTE, G. BUTTAZZO, P. SEPPECHER: *Energies with respect to a measure and applications to low dimensional structures.* Calc. Var., **5** (1997), 37–54.

28. G. BOUCHITTE, G. BUTTAZZO, P. SEPPECHER: *Shape optimization solutions via Monge-Kantorovich equation.* C. R. Acad. Sci. Paris, **324-I** (1997), 1185–1191.

29. A. BRAIDES, G. BUTTAZZO, I. FRAGALÀ: *Riemannian approximation of Finsler metrics.* Asymptotic Anal. (to appear).

30. H. BREZIS: *Liquid crystals and energy estimates for S^2-valued maps.* In "Theory and applications of liquid crystals", IMA Vol. Math. Appl. **5**, Springer-Verlag, New York (1987), 31–52.

31. H. BREZIS: *S^k-valued maps with singularities.* In "Topics in calculus of variations", Lecture Notes in Math. **1365**, Springer-Verlag, Berlin (1989), 1–30.

32. H. BREZIS, J. M. CORON, E. H. LIEB: *Harmonic maps with defects.* Commun. Math. Phys., **107** (1986), 649–705.

33. T. BRIANÇON: *Problèmes de régularité en optimisation de formes.* Ph.D thesis, ENS Cachan-Bretagne (2002).

34. F. BROCK, V. FERONE, B. KAWOHL: *A symmetry problem in the calculus of variations.* Calc. Var. Partial Differential Equations, **4** (1996), 593–599.

35. D. BUCUR: *Shape analysis for Nonsmooth Elliptic Operators.* Appl. Math. Lett., **9** (1996), 11–16.

36. D. BUCUR, G. BUTTAZZO: *Results and questions on minimum problems for eigenvalues.* Preprint Dipartimento di Matematica Università di Pisa, Pisa (1998).

37. D. BUCUR, G. BUTTAZZO: *Variational Methods in some Shape Optimization Problems.* Lecture Notes of courses at Dipartimento di Matematica Università di Pisa and Scuola Normale Superiore di Pisa, Series "Appunti della Scuola Normale Superiore", Pisa (2002).

38. D. BUCUR, G. BUTTAZZO, I. FIGUEIREDO: *On the attainable eigenvalues of the Laplace operator.* SIAM J. Math. Anal., **30** (1999), 527–536.

39. D. BUCUR, G. BUTTAZZO, A. HENROT: *Existence results for some optimal partition problems.* Adv. Math. Sci. Appl., **8** (1998), 571–579.

40. D. BUCUR, G. BUTTAZZO, P. TREBESCHI: *An existence result for optimal obstacles.* J. Funct. Anal., **162** (1999), 96–119.

41. D. BUCUR, G. BUTTAZZO, N. VARCHON: *On the problem of optimal cutting*, SIAM J. Optimization (to appear).

42. D. BUCUR, A. HENROT: *Minimization of the third eigenvlaue of the Dirichlet Laplacian*. Proc. Roy. Soc. London, Ser. A, **456** (2000), 985–996.

43. D. BUCUR, A. HENROT: *Stability for the Dirichlet problem under continuous Steiner symmetrization*. Potential Anal., **13** (2000), 127–145.

44. D. BUCUR, A. HENROT, J. SOKOLOWSKI, A. ZOCHOWSKI: *Continuity of the elasticity system solutions with respect to boundary variations*. Adv. Math. Sci. Appl., **11** (2001), 57–73.

45. D. BUCUR, P. TREBESCHI: *Shape optimization problem governed by nonlinear state equation*. Proc. Roy. Soc. Edinburgh, **128 A** (1998), 945–963.

46. D. BUCUR, N. VARCHON: *Boundary variation for the Neumann problem* , Ann. Scuola Norm. Sup. Pisa Cl. Sci., **XXIV** (2000), 807–821.

47. D. BUCUR, N. VARCHON: *A duality approach for the boundary variations of Neumann problems*, Preprint Université de Franche-Comté n. 00/14, Besançon (2000).

48. D. BUCUR, J. P. ZOLESIO: *N-Dimensional Shape Optimization under Capacitary Constraints*. J. Differential Equations, **123** (2) (1995), 504–522.

49. D. BUCUR, J. P. ZOLESIO: *Shape continuity for Dirichlet-Neumann problems*. Progress in partial differential equations: the Metz surveys, 4, Pitman Res. Notes Math. Ser. **345**, Longman, Harlow (1996), 53–65.

50. D. BUCUR, J. P. ZOLESIO *Wiener's criterion and shape continuity for the Dirichlet problem*. Boll. Un. Mat. Ital. B (7) 11 (1997), 757–771.

51. G. BUTTAZZO: *Thin insulating layers: the optimization point of view*. Proceedings of "Material Instabilities in Continuum Mechanics and Related Mathematical Problems", Edinburgh 1985–1986, edited by J. M. Ball, Oxford University Press, Oxford (1988), 11–19.

52. G. BUTTAZZO: *Semicontinuity, Relaxation and Integral Representation in the Calculus of Variations*. Pitman Res. Notes Math. Ser. **207**, Longman, Harlow (1989).

53. G. BUTTAZZO, G. DAL MASO: *Shape optimization for Dirichlet problems: relaxed solutions and optimality conditions*. Bull. Amer. Math. Soc., **23** (1990), 531–535.

54. G. BUTTAZZO, G. DAL MASO: *Shape optimization for Dirichlet problems: relaxed formulation and optimality conditions*. Appl. Math. Optim., **23** (1991), 17–49.

55. G. BUTTAZZO, G. DAL MASO: *An existence result for a class of shape optimization problems*. Arch. Rational Mech. Anal., **122** (1993), 183–195.

56. G. BUTTAZZO, G. DAL MASO, A. GARRONI, A. MALUSA: *On the relaxed formulation of Some Shape Optimization Problems*. Adv. Math. Sci. Appl., **7** (1997), 1–24.

57. G. BUTTAZZO, A. DAVINI, I. FRAGALÀ, F. MACIÁ: *Optimal Riemannian distances preventing mass transfer*. Preprint Dipartimento di Matematica Università di Pisa, Pisa (2002).

58. G. BUTTAZZO, L. DE PASCALE, I. FRAGALÀ: *Topological equivalence of some variational problems involving distances*. Discrete and Continuous Dynamical Systems, **7** (2001), 247–258.

59. G. BUTTAZZO, V. FERONE, B. KAWOHL: *Minimum problems over sets of concave functions and related questions*. Math. Nachr., **173** (1995), 71–89.

60. G. BUTTAZZO, L. FREDDI: *Relaxed optimal control problems and applications to shape optimization.* Lecture notes of a course held at the NATO-ASI Summer School "Nonlinear Analysis, Differential Equations and Control", Montreal, July 27 – August 7, 1998, Kluwer, Dordrecht (1999), 159–206.
61. G. BUTTAZZO, M. GIAQUINTA, S. HILDEBRANDT: *One-dimensional Calculus of Variations: an Introduction.* Oxford University Press, Oxford (1998).
62. G. BUTTAZZO, P. GUASONI: *Shape optimization problems over classes of convex domains.* J. Convex Anal., **4** (1997), 343–351.
63. G. BUTTAZZO, B. KAWOHL: *On Newton's problem of minimal resistance.* Math. Intelligencer, **15** (1993), 7–12.
64. G. BUTTAZZO, E. OUDET, E. STEPANOV: *Optimal transportation problems with free Dirichlet regions.* Preprint Dipartimento di Matematica Università di Pisa, Pisa (2002).
65. G. BUTTAZZO, B. SCHWEIZER: *Γ convergence of Hausdorff measures.* Preprint Dipartimento di Matematica Università di Pisa, Pisa (2002).
66. G. BUTTAZZO, P. TREBESCHI: *The role of monotonicity in some shape optimization problems.* In "Calculus of Variations, Differential Equations and Optimal Control", Research Notes in Mathematics Series, Vol. 410-411, Chapman & Hall/CRC Press, Boca Raton (1999).
67. G. BUTTAZZO, A. WAGNER: *On the optimal shape of a rigid body supported by an elastic membrane.* Nonlinear Anal., **39** (2000), 47–63.
68. G. BUTTAZZO, O. M. ZEINE: *Un problème d'optimisation de plaques.* Modél. Math. Anal. Numér., **31** (1997), 167–184.
69. E. CABIB: *A relaxed control problem for two-phase conductors.* Ann. Univ. Ferrara - Sez. VII - Sc. Mat., **33** (1987), 207–218.
70. E. CABIB, G. DAL MASO: *On a class of optimum problems in structural design.* J. Optimization Theory Appl., **56** (1988), 39–65.
71. G. CARLIER, T. LACHAND-ROBERT: *Regularity of solutions for some variational problems subject to a convexity constraint.* Comm. Pure Appl. Math., **54** (2001), 583–594.
72. J. CEA, K. MALANOWSKI: *An example of a max-min problem in partial differential equations.* SIAM J. Control, **8** (1970), p. 305–316.
73. A. CHAMBOLLE, F. DOVERI: *Continuity of Neumann linear elliptic problems on varying two-dimensional bounded open sets.*, Commun. Partial Differ. Equations, **22** (1997), 811–840.
74. A. CHAMBOLLE: *A density result in two-dimensional linearized elasticity and applications*, Preprint Ceremade, Paris (2001).
75. D. CHENAIS: *On the existence of a solution in a domain identification problem.* J. Math. Anal. Appl., **52** (1975), 189–219.
76. D. CHENAIS: *Homéomorphisme entre ouverts lipschitziens.* Ann. Mat. Pura Appl. (4) 118 (1978), 343–398.
77. M. CHIPOT, G. DAL MASO: *Relaxed shape optimization: the case of nonnegative data for the Dirichlet problem.* Adv. Math. Sci. Appl., **1** (1992), 47–81.
78. D. CIORANESCU, F. MURAT: *Un terme étrange venu d'ailleurs.* Nonlinear partial differential equations and their applications. Collège de France Seminar, Vol. II (Paris, 1979/1980), pp. 98–138, 389–390, Res. Notes in Math., 60, Pitman, Boston, Mass.-London (1982).
79. M. COMTE, T. LACHAND-ROBERT: *Existence of minimizers for Newton's problem of the body of minimal resistance under a single impact assumption.* J. Anal. Math. **83** (2001), 313–335.

80. M. COMTE, T. LACHAND-ROBERT: *Newton's problem of the body of minimal resistance under a single-impact assumption.* Calc. Var. Partial Differential Equations **12** (2001), 173–211.

81. S. J. COX: *The shape of the ideal column.* Math. Intelligencer, **14** (1992), 16–24.

82. B. DACOROGNA: *Direct Methods in the Calculus of Variations.* Appl. Math. Sciences **78**, Springer-Verlag, Berlin (1989).

83. G. DAL MASO: *An Introduction to Γ-convergence.* Birkhäuser, Boston (1993).

84. G. DAL MASO, A. DE FRANCESCHI: *Limits of nonlinear Dirichlet problems in varying domains.* Manuscripta Math., **61** (1988), 251–268.

85. G. DAL MASO, A. GARRONI: *New results on the asymptotic behaviour of Dirichlet problems in perforated domains.* Math. Mod. Meth. Appl. Sci., **3** (1994), 373–407.

86. G. DAL MASO, A. MALUSA: *Approximation of relaxed Dirichlet problems by boundary value problems in perforated domains.* Proc. Roy. Soc. Edinburgh Sect. **A-125** (1995), 99–114.

87. G. DAL MASO, F. MURAT: *Asymptotic behavior and correctors for Dirichlet problems in perforated domains with homogeneous monotone operators.* Ann. Scuola Norm. Sup. Pisa, **24** (1997), 239–290.

88. G. DAL MASO, R. TOADER: *A model for the quasi-static growth of a brittle fracture: existence and approximation results.* Preprint SISSA, Trieste (2001).

89. E. DE GIORGI: *Teoremi di semicontinuità nel calcolo delle variazioni.* Notes of a course given at the Istituto Nazionale di Alta Matematica, Rome (1968).

90. E. DE GIORGI: *Γ-convergenza e G-convergenza.* Boll. Un. Mat. Ital., **14-A** (1977), 213–224.

91. E. DE GIORGI, F. COLOMBINI, L.C. PICCININI: *Frontiere orientate di misura minima e questioni collegate.* Quaderni della Scuola Normale Superiore, Pisa (1972).

92. E. DE GIORGI, T. FRANZONI: *Su un tipo di convergenza variazionale.* Atti Accad. Naz. Lincei Cl. Sci. Fis. Mat. Natur., (8) **58** (1975), 842–850.

93. M. DELFOUR, J.-P. ZOLESIO: *Shapes and geometries. Analysis, differential calculus, and optimization.* Advances in Design and Control (SIAM), Philadelphia (2001).

94. F. DEMENGEL: *Déplacements à déformations bornées et champs de contrainte mesures.* Ann. Scuola Norm. Sup. Pisa Cl. Sci., **12** (1985), 243–318.

95. F. DEMENGEL, P. SUQUET: *On locking materials.* Acta Applicandae Math., **6** (1986), 185–211.

96. L. DE PASCALE, A. PRATELLI: *Regularity properties for Monge transport density and for solutions of some shape optimization problem.* Calc. Var., **14** (2002), 249–274.

97. L. DE PASCALE, L. C. EVANS, A. PRATELLI: *Integral estimates for transport densities.* Work in preparation.

98. I. EKELAND, R. TEMAM: *Convex Analysis and Variational Problems.* Studies in Mathematics and its Applications **1**, North-Holland, Amsterdam (1976).

99. L. C. EVANS: *Partial differential equations and Monge-Kantorovich mass transfer.* Current Developments in Mathematics, Cambridge MA (1997), 65–126, Int. Press, Boston (1999).

100. L. C. EVANS, W. GANGBO: *Differential Equations Methods for the Monge-Kantorovich Mass Transfer Problem.* Mem. Amer. Math. Soc. **137**, Providence (1999).

101. H. FEDERER: *Geometric Measure Theory.* Springer-Verlag, Berlin (1969).
102. M. FELDMAN, R.J. McCANN: *Monge's transport problem on a Riemannian manifold.* Trans. Amer. Math. Soc., **354** (2002), 1667–1697.
103. S. FINZI VITA: *Numerical shape optimization for relaxed Dirichlet problems.* Preprint Università di Roma "La Sapienza", Roma (1990).
104. S. FINZI VITA: *Constrained shape optimization for Dirichlet problems: discretization via relaxation.* Preprint Università di Roma "La Sapienza", **42** (1996).
105. G. A. FRANCFORT, F. MURAT: *Homogenization and optimal bounds in linear elasticity.* Arch. Rational Mech. Anal., **94** (1986), 307-334.
106. W. GANGBO, R. J. McCANN: *The geometry of optimal transportation.* Acta Math., **177** (1996), 113–161.
107. E. GIUSTI: *Minimal Surfaces and Functions of Bounded Variation.* Birkhäuser, Boston (1984).
108. E. GIUSTI: *Metodi diretti nel calcolo delle variazioni.* Unione Matematica Italiana, Bologna (1994).
109. F. GOLAY, P. SEPPECHER: *Locking materials and the topology of optimal shapes.* Eur. J. Mech. A Solids, **20** (4) (2001), 631–644.
110. H. H. GOLDSTINE: *A History of the Calculus of Variations from the 17th through the 19th Century.* Springer-Verlag, Heidelberg (1980).
111. P. GUASONI: *Problemi di ottimizzazione di forma su classi di insiemi convessi.* Tesi di Laurea, Università di Pisa, 1995-1996.
112. M. HAYOUNI: *Sur la minimisation de la première valeur propre du laplacien.* C. R. Acad. Sci. Paris Sér. I Math., **330** (2000), 551–556.
113. A. HENROT: *Continuity with respect to the domain for the Laplacian: a survey.* Control and Cybernetics, **23** (1994), 427–443.
114. A. HENROT, M. PIERRE: *Optimisation de forme* (book in preparation).
115. D. HORSTMANN, B. KAWOHL, P. VILLAGGIO: *Newton's aerodynamic problem in the presence of friction.* Preprint University of Cologne, Cologne (2000).
116. B. KAWOHL, L. TARTAR, O. PIRONNEAU, J.-P. ZOLESIO: *Optimal Shape Design.* Springer-Verlag, Berlin, 2001.
117. R. V. KOHN, G. STRANG: *Optimal design and relaxation of variational problems, I,II,III.* Comm. Pure Appl. Math., **39** (1986), 113–137, 139–182, 353–377.
118. R. V. KOHN, M. VOGELIUS: *Relaxation of a variational method for impedance computed tomography.* Comm. Pure Appl. Math., **40** (1987), 745–777.
119. T. LACHAND-ROBERT, M.A. PELETIER: *An example of non-convex minimization and an application to Newton's problem of the body of least resistance.* Ann. Inst. H. Poincaré Anal. Non Linéaire, **18** (2001), 179–198
120. T. LACHAND-ROBERT, M.A. PELETIER: *Newton's problem of the body of minimal resistance in the class of convex developable functions.* Math. Nachr., **226** (2001), 153–176.
121. E. H. LIEB: *On the lowest eigenvalue of the Laplacian for the intersection of two domains.* Invent. Math., **74**, (1983), 441–448.
122. P. L. LIONS: *The concentration-compactness principle in the Calculus of Variations. The locally compact case, part 1.* Ann. Inst. Poincaré, **1**, (1984), 109–145.
123. W. LIU, P. NEITTAANMAKI, D. TIBA: *Sur les problèmes d'optimisation structurelle.* C. R. Acad. Sci. Paris, **I-331** (2000), 101–106.

124. K. A. LURIE, A. V. CHERKAEV: *G-closure of a Set of Anisotropically Conductivity Media in the Two-Dimensional Case.* J. Optimization Theory Appl., **42** (1984), 283–304.

125. K. A. LURIE, A. V. CHERKAEV: *G-closure of Some Particular Sets of Admissible Material Characteristics for the Problem of Bending of Thin Elastic Plates.* J. Optimization Theory Appl., **42** (1984), 305–316.

126. P. MARCELLINI: *Nonconvex integrals of the calculus of variations.* Methods of nonconvex analysis (Varenna, 1989), 16–57, Lecture Notes in Math. **1446**, Springer-Verlag, Berlin (1990).

127. V. G. MAZ'JA: *Sobolev Spaces.* Springer-Verlag, Berlin (1985).

128. R. J. McCANN: *Existence and uniqueness of monotone measure-preserving maps.* Duke Math. J., **80** (1995), 309–323.

129. A. MIELE: *Theory of Optimum Aerodynamic Shapes.* Academic Press, New York (1965).

130. G. MONGE: *Memoire sur la Theorie des Déblais et des Remblais.* Historie de l'Académie Royale des Sciences de Paris, avec les Mémoires de Mathématique et de Physique pour la Même année, (1781), 666–704.

131. F. MORGAN: *Geometric Measure Theory, a Beginners Guide.* Academic Press, New York (1988).

132. F. MORGAN, R. BOLTON: *Hexagonal economic regions solve the location problem.* Amer. Math. Monthly, **109** (2001), 165–172.

133. C. B. MORREY: *Multiple integrals in the calculus of variations.* Springer, Berlin (1966).

134. U. MOSCO: *Convergence of convex sets and of solutions of variational inequalities.* Adv. in Math., **3** (1969), 510–585.

135. U. MOSCO: *Composite media and asymptotic Dirichlet forms.* J. Funct. Anal., **123** (1994), 368–421.

136. F. MURAT, J. SIMON: *Sur le contrôle par un domaine géometrique.* Preprint 76015, Univ. Paris VI, (1976).

137. F. MURAT, L. TARTAR: *Calcul des variations et homogénéisation.* Proceedings of "Les Méthodes de l'homogénéisation: Théorie et applications en physique", Ecole d'Eté d'Analyse Numérique C.E.A.-E.D.F.-INRIA, Bréausans-Nappe 1983, Collection de la direction des études et recherches d'electricité de France **57**, Eyrolles, Paris, (1985), 319–369.

138. F. MURAT, L. TARTAR: *Optimality conditions and homogenization.* Proceedings of "Nonlinear variational problems", Isola d'Elba 1983, Res. Notes in Math. **127**, Pitman, London, (1985), 1–8.

139. P. NEITTAANMAKI, D. TIBA: *Shape optimization in free boundary systems.* Free boundary problems: theory and applications, II (Chiba, 1999), 334–343, GAKUTO Internat. Ser. Math. Sci. Appl., **14**, Gakkōtosho, Tokyo (2000).

140. F. OTTO: *The geometry of dissipative evolution equations: the porous medium equation.* Comm. Partial Differential Equations, **26** (2001), 101–174.

141. E. OUDET: *Shape Optimization and Control.* Ph.D thesis, ULP Strasbourg, France. In preparation.

142. O. PIRONNEAU: *Optimal Shape Design for Elliptic Systems.* Springer-Verlag, Berlin (1984).

143. G. POLYA: *On the characteristic frequencies of a symmetric membrane,* Math. Zeit., **63** (1955), 331–337.

144. S. T. RACHEV, L. RÜSCHENDORF: *Mass transportation problems. Vol. I Theory, Vol. II Applications*. Probability and its Applications, Springer-Verlag, Berlin (1998).
145. J. SOKOLOWSKI, A. ZOCHOVSKI: *On the topological derivative in shape optimization*. SIAM J. Control Optim., **37** (1999), 1251–1272
146. J. SOKOLOWSKI, J.-P. ZOLESIO: *Introduction to shape optimization. Shape sensitivity analysis*. Springer Series in Computational Mathematics, 16. Springer-Verlag, Berlin (1992).
147. V. ŠVERÁK: *On optimal shape design*. J. Math. Pures Appl., **72** (1993), 537–551.
148. L. TARTAR: *Estimations Fines des Coefficients Homogénéises*. Ennio De Giorgi Colloquium, Edited by P.Krée, Res. Notes in Math. **125**, Pitman, London (1985), 168–187.
149. L. TARTAR: *An introduction to the homogenization method in optimal design*. In "Optimal Shape Design", Lecture Notes in Math. **1740**, Springer-Verlag, Berlin (2000), 47–156.
150. R. TOADER: *Wave equation in domains with many small obstacles*. Asymptot. Anal., **23** (2000), 273–290.
151. J. URBAS: *Mass transfer problems*. Unpublished manuscript.
152. N. VAN GOETHEM: *Variational problems on classes of convex domains*. Preprint Dipartimento di Matematica Università di Pisa, Pisa (2000).
153. S. VENTURINI: *Derivation of distance functions in $\mathbb{R}^N$*. Preprint Dipartimento di Matematica, Università di Bologna (1991).
154. C. VILLANI: *Topics in mass transportation*. Book in preparation.
155. A. WAGNER: *A remark on Newton's resistance formula*. Preprint University of Cologne, Cologne (1998).
156. S. A. WOLF, J. B. KELLER: *Range of the first two eigenvalues of the Laplacian*. Proc. Roy. Soc. Lond., **A-447** (1994), 397–412.

Optimal transportation, dissipative PDE's and functional inequalities

Cedric Villani

UMPA, École Normale Supérieure de Lyon,
F-69364 Lyon Cedex 07, FRANCE.
cvillani@umpa.ens-lyon.fr

Recent research has shown the emergence of an intricate pattern of tight links between certain classes of optimal transportation problems, certain classes of evolution PDE's and certain classes of functional inequalities. It is my purpose in these notes to convey an idea of these links through (hopefully) pedagogical examples taken from recent works by various authors. During this process, we shall encounter such diverse areas as fluid mechanics, granular material physics, mean-field limits in statistical mechanics, and optimal Sobolev inequalities.

I have written two other texts dealing with mass transportation techniques, which may complement the present set of notes. One [41] is a set of lectures notes for a graduate course taught in Georgia Tech, Atlanta; the other one [40] is a short contribution to the proceedings of a summer school in the Azores, organized by Maria Carvalho; I have tried to avoid repetition. With respect to both abovementioned sources, the present notes aim at giving a more impressionist picture, with priority given to the diversity of applications rather than to the systematic nature of the exposition. The plan here is the opposite of the one that you would expect in a course: it starts with applications and ends up with theoretical background. There is a lot of overlapping with the proceedings of the Azores summer school, however the latter was mainly focusing on the problem of trend to equilibrium for dissipative equations. Most of the material in sections III and IV is absent from the Atlanta lecture notes.

I chose not to start by giving precise definitions of mass transportation; in fact, each lecture will present a slightly different view on mass transportation.

It is a pleasure to thank the CIME organization for their beautiful work on the occasion of the summer school in Martina Franca, in which these lectures have been given. I also thank Yann Brenier for his suggestions during the preparation of these lectures.

Contents

1 Some motivations

Let me start by stating several problems which seem loosely related, and do not seem to have anything to do with mass transportation.

Problem # 1 (rate of convergence as $t \to \infty$): Consider the diffusive, nonlinear partial differential equation where the unknown $(f_t)_{t \geq 0}$ is a time-dependent probability density on $\mathbb{R}^d$,

$$\frac{\partial f}{\partial t} = \sigma \Delta f + \nabla \cdot (f \nabla V) + \nabla \cdot (f \nabla (f * W)), \qquad t \geq 0, \; x \in \mathbb{R}^d \qquad (1)$$

(here ∇ stands for the gradient operator in $\mathbb{R}^d$, while $\nabla \cdot$ is its adjoint, the divergence operator; moreover, V and W are smooth potentials). Does this equation admit a stationary state ? If the answer is yes, do solutions of (1) converge towards this stationary state ?

Problem # 2 (rate of convergence as $N \to \infty$): Consider a bunch of N particles in $\mathbb{R}^d$, with respective positions X_t^i ($1 \leq i \leq N$) at time $t \geq 0$, solutions of the stochastic differential equation

$$dX_t^i = dB_t^i - \nabla V(X_t^i) \, dt - \frac{1}{N} \sum_{j=1}^{N} \nabla W(X_t^i - X_t^j) \, dt, \qquad 1 \leq i \leq N,$$

starting from some chaotic initial configuration $X_0 = (X_0^1, \dots, X_0^N)$ with law $\mathcal{L}(X_0) = f_0^{\otimes N} \, dx$ on $(\mathbb{R}^d)^N$. Under some assumptions on V and W it is known that for each time $t > 0$ the density of particles in $\mathbb{R}^d$ associated with this system converges, as $N \to \infty$, towards the unique solution of (1). This means for instance, that for any bounded continuous function φ,

$$\frac{1}{N} \sum_{i=1}^{N} \varphi(X_t^i) \xrightarrow[N \to \infty]{} \int_{\mathbb{R}^d} f_t \, \varphi \, dx,$$

in the sense of convergence in law for random variables (note that the left-hand side is random, while the right-hand side is not). Can one estimate the speed of convergence as $N \to \infty$?

Problem # 3 (optimal constants): What is the optimal constant, and how do minimizers look like, in the Gagliardo-Nirenberg interpolation inequality

$$\|w\|_{L^p(\mathbb{R}^n)} \leq C\|\nabla w\|_{L^2(\mathbb{R}^n)}^{\theta}\|w\|_{L^q(\mathbb{R}^n)}^{1-\theta} \tag{2}$$

(with some compatibility conditions on p, q) ? What about the Young inequality

$$\|f * g\|_{L^r(\mathbb{R}^n)} \leq C\|f\|_{L^p(\mathbb{R}^n)}\|g\|_{L^q(\mathbb{R}^n)} \qquad ?$$

All three problems have been studied by various authors and solved in certain cases. Some of them are quite old, like the study of optimal Young inequalities, which goes back to the seventies. Among recent works, let me mention Carrillo, McCann and Villani [12] for problem 1; Malrieu [24] for problem 2; Dolbeault and del Pino [19] for the Gagliardo-Nirenberg inequality in problem 3; Barthe [4] for the Young inequality in the same problem. It turns out that in all these cases, either optimal mass transportation was explicit in the solution, or it has been found to provide much more transparent proofs than those which have been first suggested. My goal here is not to explain *why* mass transportation is efficient in such problems, but to give an idea of *how* it can be used in these various contexts.

2 A study of fast trend to equilibrium

Let us start with the study of equation (1). We shall only consider a particular case of it :

$$\frac{\partial f}{\partial t} = \Delta f + \nabla \cdot (f\nabla(f * W)), \tag{3}$$

where W is a strictly convex, symmetric $(W(-z) = W(z))$ potential, growing superquadratically at infinity. Note that equation (3) has two conservation laws, the total mass of particles and the center of mass, $\int x f(x)\,dx$. The particular case we have in mind is $W(z) = |z|^3/3$, which appears (in dimension 1) in a kinetic modelling of granular material undergoing diffusion in velocity space [5]; the case without diffusion was studied by McNamara and Young [28], Benedetto et al. [6]. On physical grounds, one could also include a term like $\theta\nabla \cdot (fx)$ $(\theta > 0)$ in the right-hand side of (3); this would in fact simplify the analysis below. Also, as we shall see the particular form of W does not really matter for us; what matters is its strict convexity.

Of course equation (3) is a diffusion equation, with a nonlinear transport term. However, this classification overlooks the fact that it has a particular structure: it can be rewritten as

$$\frac{\partial f}{\partial t} = \nabla \cdot \left(f[\nabla \log f + \nabla(f * W)] \right) \tag{4}$$

$$= \nabla \cdot \left(f \nabla \frac{\delta E}{\delta f} \right), \tag{5}$$

where

$$E(f) = \int f \log f + \frac{1}{2} \int_{\mathbb{R}^d \times \mathbb{R}^d} f(x) f(y) \, W(x - y) \, dx \, dy,$$

and $\delta E / \delta f$ stands for the gradient of E with respect to the standard L^2 Hilbert structure. In the sequel, we shall call E the "free energy" (meaning that it is some combination of an internal energy, or entropy term, and of an interaction energy).

As a consequence of (5), we immediately see that, if we set rigorous justification aside,

$$\frac{d}{dt} E(f_t) = \int \frac{\delta E}{\delta f} \nabla \cdot \left(f \nabla \frac{\delta E}{\delta f} \right) = - \int f \left| \nabla \frac{\delta E}{\delta f} \right|^2 \le 0.$$

It is therefore natural to expect that $f(t, \cdot) = f_t$ will converge towards the minimizer f_∞ of the free energy E. Here we implicitly assume the existence and uniqueness of this minimizer (in the class of probability measures with fixed center of mass), which was proven by McCann [27].

Using a standard convexity method, which happens to work in this context only for the particular case $d = 1$ and for a quadratic or cubic interaction, Benedetto et al [5] proved convergence of f_t towards f_∞ as $t \to \infty$. No rate of convergence was available.

Then mass transportation was input in the problem, via some ideas going back to Otto, and led to spectacular improvements. In a recent work, Carrillo, McCann and Villani [12] prove that the convergence is exponential :

$$\|f_t - f_\infty\|_{L^1} \le C e^{-\lambda t}, \tag{6}$$

and show how to recover explicit estimates for C and λ in terms of $\int f_0(x) |x|^2 \, dx$ only. Moreover, the proof does not need the cubic form of the potential, neither the restriction to dimension 1. It is also a "fully nonlinear" argument, in the sense that it does not use any linearization procedure (which would usually result in the destruction of any hope of estimating C).

Let us give a precise statement, slightly more general.

Theorem 2.1. *Let $f = (f_t)_{t \ge 0}$ be a solution of*

$$\frac{\partial f}{\partial t} = \Delta f + \nabla \cdot (f \nabla(f * W)),$$

where W is a C^2 symmetric interaction potential satisfying $D^2 W(z) \ge K|z|^\gamma$ for some $\gamma > 0$, and $|\nabla W(z)| \le C(1 + |z|)^\beta$ for some $\beta \ge 0$ (example : $W(z) = |z|^3 / 3$). Let

$$E(f) = \int f \log f + \frac{1}{2} \int W(x-y) \, f(x) \, f(y) \, dx \, dy,$$

and let f_∞ be the unique minimizer of F with the same center of mass as f. Then, for any $t_0 > 0$ there exist constants $C_0, \lambda_0 > 0$, explicitly computable and depending on f only via an upper bound for $\int f_0(x)|x|^2 \, dx$, such that

$$t \geq t_0 \implies \|f_t - f_\infty\|_{L^1} \leq C_0 e^{-\lambda_0 t}.$$

How does the argument work ? The first main idea (quite natural) is to focus on the rate of dissipation of energy,

$$D(f) = \int f \left| \nabla \frac{\delta E}{\delta f} \right|^2$$

$$= \int f \, |\nabla (\log f + W * f)|^2,$$

and to look for a **functional inequality** of the form

$$D(f) \geq \text{const.}[E(f) - E(f_\infty)]. \tag{7}$$

Of course, if this inequality holds true, by combining it with the identity $(d/dt)[E(f_t) - E(f_\infty)] = -D(f_t)$ we shall obtain exponential convergence, as desired. An important feature here is that we have temporarily left the world of PDE's to enter that of functional inequalities, and this will allow a lot of flexibility in the treatment.

To better understand the nature of (7), let us imagine for a few moments that there is a linear drift term $\nabla(f \nabla V)$ instead of the nonlinear term $\nabla(f \nabla(W * f))$; and accordingly, replace the interaction part of the free energy by $\int f V$. Then, assuming that V has been normalized in such a way that $\int e^{-V} = 1$, inequality (7) turns into

$$\int f |\nabla (\log f) + \nabla V|^2 \geq \text{const.} \left(\int f \log f + \int f V \right). \tag{8}$$

This inequality is well-known, and called a **logarithmic Sobolev inequality** for the reference measure e^{-V}. The terminology logarithmic Sobolev inequality may sound strange, but it is justified by the fact that (8) can be rewritten in the equivalent formulation

$$\int h^2 \, \log(h^2) \, d\mu \leq \text{const.} \int |\nabla h|^2 \, d\mu + \left(\int h^2 \, d\mu \right) \log \left(\int h^2 \, d\mu \right),$$

with $d\mu = e^{-V} \, dx$, and this last inequality asserts the embedding $H^1(d\mu) \subset L^2 \log L(d\mu)$, a weak, limit case of Sobolev embedding. This embedding is actually **not always** true; it depends on μ. A famous result by Bakry and Emery [3] states that this is indeed the case if V is uniformly convex.

Logarithmic Sobolev inequalities have become a field of mathematics by themselves, after the famous works by Gross in the mid-seventies. The interesting reader can find many references in the surveys [20] (already 10 years old) and [1] (in french). This may seem strange, because they would seem to be weaker than usual Sobolev inequalities. One reason why they are so important to many researchers, is that the constants involved do not depend on the dimension of space, contrarily to standard Sobolev inequalities; thus logarithmic Sobolev inequalities are a substitute to Sobolev inequalities in infinite dimension. As far as we are concerned, they have another interest: they are well-adapted to the study of the asymptotic behavior of diffusion equations, because they behave well at the neighborhood of the minimizer of E (both sides of (8) vanish simultaneously). In view of this, it is natural to look upon (7) as a "nonlinear" generalization of (8).

There are many known proofs of (8), but the great majority of them fail to yield generalizations like (7). However, at the moment we know of two possible ways towards (7). One is the adaptation of the so-called Bakry-Emery strategy, and has been obtained by a reinterpretation of a famous argument from [3]. It is based on computing the *second derivative* of $E(f_t)$ with respect to time, and showing that the resulting functional, call it $DD(f)$, can be compared directly to $D(f)$; then use the evolution equation to convert this result of comparison between DD and D, into a result of comparison between D and E. The other strategy relies on mass transportation, and more precisely the displacement interpolation technique introduced in McCann [27] and further investigated in Otto [32], Otto and Villani [33]. We insist that even if mass transportation is absent from the first argument, it can be seen as underlying it, as explained in [33] or [12].

In the sequel we only explain about the second strategy; we note that, at least in some cases, its implementation can be significantly simplified by a technique due to Cordero-Erausquin [15, 16]. In this point of view, the key property behind (7) is **displacement convexity**. Let us explain this simple, but quite interesting concept.

Let f_0 and f_1 be two probability densities on $\mathbb{R}^d$. To be more intrinsic we should state everything in terms of probability *measures*, but in the present context it is possible to deal with functions. By a theorem of Brenier [9] and McCann [26] and maybe earlier authors, there exists a unique gradient of convex function (unique on the support of f_0), $\nabla\varphi$, which **transports** f_0 onto f_1, in the sense that the image measure of $f_0(x)\,dx$ by $\nabla\varphi$ is $f_1(x)\,dx$. In other words, for all bounded continuous function h on $\mathbb{R}^d$,

$$\int h(x)f_1(x)\,dx = \int h(\nabla\varphi(x))f_0(x)\,dx.$$

We shall use the notation

$$\nabla\varphi\#f_0 = f_1.$$

Moreover, this map $T = \nabla\varphi$ is obtained as the unique solution of **Monge's optimal mass transportation problem**

$$\inf_{T\#f_0=f_1} \int f_0(x)|x - T(x)|^2 \, dx.$$

Note that this construction naturally introduces a notion of distance between f_0 and f_1. This notion coincides with the Monge-Kantorovich (or Wasserstein) distance of order 2,

$$W_2(f_0, f_1) = \sqrt{\inf_{T\#f_0=f_1} \int f_0(x)|x - T(x)|^2 \, dx}$$

$$= \sqrt{\int f_0(x)|x - \nabla\varphi(x)|^2 \, dx}.$$

As noticed by McCann [27], this procedure makes it possible to define a natural "interpolant" $(f_s)_{0\leq t\leq 1}$ between f_0 and f_1, by

$$f_s = [(1 - s)\mathrm{id} + s\nabla\varphi]\#f_0.$$

By definition, a functional E is displacement convex if, whenever f_0 and f_1 are two probability densities, the function $E(f_s)$ is a convex function of the parameter s. To compare this definition with that of convexity in the usual sense, note that the latter can be rephrased as: *for all f_0 and f_1, the function $E((1 - s)f_0 + sf_1)$ is a convex function of $s \in [0, 1]$.*

Just as for usual convexity, one can refine the concept of displacement convexity into that of λ-uniform displacement convexity ($\lambda \geq 0$), meaning that

$$\frac{d^2}{ds^2}E(f_s) \geq \lambda W_2(f_0, f_1)^2.$$

Of course 0-uniform displacement convexity is just displacement convexity.

At this point it is useful to give some examples. Researchers working on these questions have focused on three model functionals :

- $E(f) = \int fV$ is λ-displacement convex if $D^2V \geq \lambda\mathrm{Id}$. Here D^2 stands for the Hessian operator on $C^2(\mathbb{R}^d)$.

- $E(f) = \frac{1}{2}\int f(x)f(y)W(x - y) \, dx \, dy$ is λ-uniformly displacement convex if $D^2W \geq \lambda\mathrm{Id}$, and if one restricts to some set of probability measures with fixed center of mass ($\int f_0(x)x \, dx = \int f_1(x)x \, dx$ in the definition). Note that in our context, W is not in general uniformly convex, so that we cannot go beyond displacement convexity.

- $E(f) = \int U(f(x)) \, dx$ is displacement convex if $r \mapsto r^dU(r^{-d})$ is convex nonincreasing on $\mathbb{R}_+$. This is in particular the case if $U(r) = r\log r$, as in the present situation.

Now comes the core of the argument towards (7). It consists in using displacement convexity to establish the stronger functional inequality, holding for all probability densities f_0 and f_1 with common center of mass,

$$E(f_0) - E(f_1) \leq \sqrt{D(f_0)} W_2(f_0, f_1) - \frac{K}{2} W(f_0, f_1)^2, \qquad (9)$$

for some constant $K > 0$, depending on f_0 and f_1 only via an upper bound on $E(f_0)$ and $E(f_1)$.

How should one think of inequality (9)? If we see things from a PDE point of view, the left-hand side measures some discrepancy between f_0 and f_1 in some sense which looks slightly stronger than L^1 (something like $L \log L$). On the other hand, the right-hand side involves both a weak control of the distance between f_0 and f_1 (the W_2 terms) and a control of the smoothness of f_0 (the $D(f_0)$ term, which involves gradients of f_0). It can therefore be thought of as an **interpolation inequality**.

The interest of (9) for our problem is clear: by Young's inequality, it implies

$$E(f_0) - E(f_1) \leq \frac{1}{2K} D(f_0),$$

which is precisely what we are looking at if we replace f_1 by the minimizer f_∞ of E. The constant K will only depend on an upper bound on the entropy of $E(f_0)$; we will see later how this assumption can be dispended of in the final results.

On the other hand, as we shall see in a moment, (9) can be proved rather easily by displacement convexity, based on the Taylor formula

$$E(f_1) = E(f_0) + \frac{d}{ds}\Big|_{s=0} E(f_s) + \int_0^1 (1-s) \frac{d^2}{ds^2} E(f_s)\, ds, \qquad (10)$$

and the estimates

$$\frac{d^2}{ds^2} E(f_s) \geq K W_2(f_0, f_1)^2, \qquad (11)$$

$$\left| \frac{d}{ds}\Big|_{s=0} E(f_s) \right| \leq \sqrt{D(f_0)} W_2(f_0, f_1). \qquad (12)$$

Clearly, the combination of (10), (11) and (12) solves our problem. We mention that the very same strategy was used in Otto and Villani [33] to give a new proof of the Bakry-Emery theorem.

Let us first show how to prove (11) in the case when W is uniformly convex, i.e. there exists some $\lambda > 0$ such that $D^2 W \geq \lambda \mathrm{Id}$. First let us show what was claimed above, namely that in this situation E is λ-uniformly displacement convex when the center of mass is fixed. Let f_0 and f_1 be two probability densities, and $(f_s)_{0 \leq s \leq 1}$ the associated displacement interpolant. By the change of variable formula, f_s satisfies the Monge-Ampère type equation

$$f_0(x) = f_s\big((1-s)x + s\nabla\varphi(x)\big)\det\big((1-s)\mathrm{Id} + sD^2\varphi\big), \tag{13}$$

On the other hand, also by changing variables, we have

$$\int f_s \log f_s =$$

$$\int f_s\big((1-s)x + s\nabla\varphi(x)\big)\log f_s\big((1-s)x + s\nabla\varphi(x)\big)\det\big((1-s)\mathrm{Id} + sD^2\varphi\big),$$

and combining this with (13), one finds

$$\int f_s \log f_s = \int f_0(x)\log\frac{f_0(x)}{\det\big((1-s)\mathrm{Id} + sD^2\varphi\big)},$$

$$= \int f_0 \log f_0 - \int f_0 \log\det\big((1-s)\mathrm{Id} + sD^2\varphi\big). \tag{14}$$

A nontrivial result by McCann [27] shows that this procedure is rigorous. Then, one sees that the right-hand side is a convex function of s.

Next, by the definition of image measure,

$$\frac{1}{2}\int f_s(x)f_s(y)W(x-y)\,dx\,dy$$

$$= \frac{1}{2}\int f_0(x)f_0(y)W\big([(1-s)x + s\nabla\varphi(x)] + [(1-s)y + s\nabla\varphi(y)]\big)\,dx\,dy, \tag{15}$$

and from this we see that

$$\frac{d^2}{ds^2}\frac{1}{2}\int f_s(x)f_s(y)W(x-y)\,dx\,dy = \frac{1}{2}\int f_0(x)f_0(y)\Big\langle D^2W\big((1-s)(x-y)$$

$$+ s(\nabla\varphi(x) - \nabla\varphi(y))\big)\cdot[\xi(x) - \xi(y)], [\xi(x) - \xi(y)]\Big\rangle\,dx\,dy,$$

where we use the shorthand $\xi(x) = x - \nabla\varphi(x)$. ¿From our assumption on D^2W, this is bounded below by

$$\frac{\lambda}{2}\int f_0(x)f_0(y)|\xi(x) - \xi(y)|^2\,dx\,dy,$$

which turns out to be precisely

$$2\frac{\lambda}{2}\int f_0(x)|\xi(x)|^2\,dx = \lambda W_2(f_0, f_1)^2,$$

in view of the identity

$$\int f_0(x)\xi(x)\,dx = \int f_0(x)x\,dx - \int f_0(x)\nabla\varphi(x)\,dx$$

$$= \int f_0(x)x\,dx - \int f_1(y)y\,dy = 0.$$

To summarize, at this point we have shown (11) under the restrictive assumption that W be uniformly convex. What if W behaves like the cubic potential? For instance, let us assume $D^2W(z) \geq \psi(|z|)$, where ψ is continuous and vanishes at $z = 0$. Then we can treat this case just as the previous one, with the help of the following unexpected lemma (certainly suboptimal):

$$\int f_0(x)f_0(y)\psi(|x-y|)|\xi(x) - \xi(y)|^2\,dx\,dy \geq$$

$$K(f_0)\int f_0(x)f_0(y)|\xi(x) - \xi(y)|^2\,dx\,dy,$$

$$K(f_0) = \frac{1}{8}\inf_{z_1,z_2 \in \mathbb{R}^d}\int f_0(x)\inf[\psi(|x-z_1|), \psi(|x-z_2|)]\,dx.$$

Note that $K(f_0)$ is strictly positive as soon as f_0 is a probability density; and can be estimated from below in terms of an upper bound on $E(f_0)$.

Now, let us sketch the proof of (12). From (14) and (15), it is not difficult to compute explicitly

$$\frac{d}{ds}\bigg|_{s=0} E(f_s) = -\int f_0(x)[\Delta\varphi(x) - d] + \int f_0(x)f_0(y)\nabla W(x-y) \cdot [\nabla\varphi(x) - x],$$

and after a few manipulations this can be rewritten as

$$\int f_0(x)\nabla(\log f_0 + f_0 * W)(x) \cdot [\nabla\varphi(x) - x]\,dx.$$

Then, by Cauchy-Schwarz inequality,

$$\left|\frac{d}{ds}\bigg|_{s=0} E(f_s)\right| \leq \sqrt{\int f_0|\nabla(\log f_0 + f_0 * W)|^2}\sqrt{\int f_0(x)|\nabla\varphi(x) - x|^2\,dx},$$

which is precisely (12). This ends our study of (7).

Let us conclude this section by explaining how one deduces (6) from (7). This will be the occasion to enter back into the PDE world... We shall just sketch the main steps, without entering their proof.

First, as a consequence of (5) and of the displacement convexity of E, one can prove [34] the estimate

$$E(f_t) \leq \frac{W_2(f_0, f_\infty)^2}{4t} + E(f_\infty). \tag{16}$$

Note that one can interpretate this inequality as a parabolic regularization inequality: it states that the size of f_t in $L \log L$ (measured by $E(f_t)$) becomes finite like $O(1/t)$ as t becomes positive.

Since $E(f_{t_0})$ is finite for any $t_0 > 0$, and since $E(f_t)$ is uniformly bounded by $E(f_{t_0})$ for $t \geq t_0$, it is then possible, for $t \geq t_0$, to apply (7) and deduce that

$$\frac{d}{dt}[E(f_t) - E(f_\infty)] \leq -\text{const.}[E(f_t) - E(f_\infty)].$$

¿From this one of course concludes that

$$E(f_t) - E(f_\infty) = O(e^{-\mu t})$$

for some $\mu > 0$.

Next, another functional inequality due to Carrillo, McCann and Villani [12] states that

$$W_2(f, f_\infty) \leq \sqrt{\text{const.}[E(f) - E(f_\infty)]}.$$

In the case when the interaction energy is replaced by the simpler energy $\int fV$, with a uniformly convex V, then there is also a similar inequality, proven by Otto and Villani [33], which is a generalization of a well-known inequality by Talagrand [36]. At this point we are able to conclude that

$$W_2(f_t, f_\infty) = O(e^{-\frac{\mu}{2}t}). \tag{17}$$

The game now is to interpolate this information with adequate smoothness bounds, to gain convergence of $\|f_t - f_\infty\|_{L^1}$. For this, one can first prove that

$$D(f_t) \leq \frac{W_2(f_t, f_\infty)^2}{t^2},$$

exactly in the same spirit as (16). ¿From this follows a uniform bound on $D(f_t)$ for $t \geq t_0 > 0$. Combining this with the easy inequality

$$\int f \left| \nabla \log \frac{f}{f_\infty} \right|^2 \leq C[D(f) + E(f)],$$

one finds that

$$\sup_{t \geq t_0} \int f_t \left| \nabla \log \frac{f_t}{f_\infty} \right|^2 < +\infty. \tag{18}$$

Estimate (18) is precisely what will play the role of smoothness bound for f_t. From the convexity of $-\log f_\infty$, one can prove an interpolation inequality similar to (9),

$$\int f \log \frac{f}{f_\infty} \le C W_2(f, f_\infty) \sqrt{\int f_t \left| \nabla \log \frac{f_t}{f_\infty} \right|^2}.$$

Combining this inequality with (17) and (18), we obtain

$$\int f_t \log \frac{f_t}{f_\infty} = O(e^{-\frac{\mu}{2}t}).$$

To conclude the argument it suffices to recall the classical Csiszár-Kullback-Pinsker inequality, in the form

$$\|f - f_\infty\|_{L^1} \le \sqrt{2 \int f \log \frac{f}{f_\infty}}.$$

3 A study of slow trend to equilibrium

In the previous section, I have explained how to use some of the mass transportation formalism in order to prove fast trend to equilibrium. In this section, I will present another example in which mass transportation enables one to prove that convergence to equilibrium has to be slow, in some sense. Moreover, we shall see that mass transportation here has a remarkable role of helping physical intuition. All of the material in this section is taken from a recent work by Caglioti and the author.

Let us start with the general definition of **Monge-Kantorovich distances**. Let $d(x, y) = |x - y|$ be the Euclidean distance (or any other continuous distance) on $\mathbb{R}^d$, and let $p \ge 1$ be a real number. Whenever μ and ν are two probability measures on $\mathbb{R}^d$, one can define their Monge-Kantorovich, or Wasserstein, distance of order p,

$$W_p(\mu, \nu) = \inf \left\{ \left[\int d(x, y)^p \, d\pi(x, y) \right]^{1/p} ; \quad \pi \in \Pi(\mu, \nu) \right\},$$

where $\Pi(\mu, \nu)$ stands for the set of all probability measures on $\mathbb{R}^d \times \mathbb{R}^d$ with marginals μ and ν. More explicitly, $\pi \in \Pi(\mu, \nu)$ if and only if, whenever φ, ψ are continuous bounded functions on $\mathbb{R}^d$,

$$\int [\varphi(x) + \psi(y)] \, d\pi(x, y) = \int \varphi \, d\mu + \int \psi \, d\nu.$$

The Monge-Kantorovich distances have been used in various contexts in statistical mechanics, going back at least to the works of Dobrushin, and Tanaka, in the seventies. Tanaka [37, 29, 38] noticed that W_2 is a *nonexpansive* distance along solutions of the spatially homogeneous Boltzmnan equation for Maxwell molecules (see [41] for a review). Even if this is a very special case of Boltzmann equation, his remark led to interesting progress and a better understanding of trend to equilibrium in this context.

Here I shall explain about another use of the Wasserstein distances in kinetic theory: for a very simple model of granular media, closely related to the example of the previous section.

First, some very sketchy background about granular media. This field has become extremely popular over the last decades, and there are dozens of models for it. Some people use kinetic models (in phase space, i.e. position and velocity) based on **inelastic** collisions, i.e. allowing deperdition of energy. In particular, inelastic Boltzmann or inelastic Enskog equations are used for this. From a mathematical point of view, these are extremely complicated models, about which essentially nothing is known (see the references in [14] and in [39]). Among basic PDE's from classical fluid mechanics, the standard Boltzmann equation is usually considered as a terribly difficult one, but the inelastic one seems to be even ways ahead in this respect.

In particular, the asymptotic behavior of solutions to the inelastic Boltzmann equation constitutes a formidable problem. There is no H theorem in this setting, and natural asymptotic states would have zero temperature, i.e. all the particles at a given position should travel at the same speed. In particular, the corresponding density of particles would be very singular. It is unknown whether concentration phenomena may occur in finite time, and it is debated whether there is a relevant hydrodynamic scaling. A key element of this debate seems to be the existence of "**homogeneous cooling states**", i.e. particular spatially homogeneous, self-similar solutions converging to a dirac mass as $t \to \infty$, taking the form

$$f_S(t, v) = \frac{1}{\alpha^N(t)} F\left(\frac{v - v_0}{\alpha(t)}\right).$$

It is believed by some authors that such states do exist and "attract" (locally in x) all solutions. Thus they would play the same role in this context, than local thermodynamical equilibria (Maxwellian distributions) do in the classical setting.

¿From a mathematical point of view, what can be said about homogeneous cooling states ? We begin with some bad news: Bobylev, Carrillo and Gamba [8] have shown, for a simplified model ("inelastic Maxwell molecules") that there does not in general exist homogeneous cooling states. Only in some restricted, weaker meaning can the concept of universal asymptotic profile be salvaged.

In this lecture we consider a *considerably* simplified model, introduced by McNamara and Young [28]. It is one-dimensional and spatially homogeneous; so the unknown is a time-dependent probability measure on $\mathbb{R}$, to be thought of as a velocity space. It reads

$$\frac{\partial f}{\partial t} = \frac{\partial}{\partial x}\left(f \frac{\partial}{\partial x}(f * W)\right), \qquad W(z) = \frac{|z|^3}{3}. \tag{19}$$

In the sequel, we shall often write $f_t(x)$ for the solution of (19), even if this solution should be thought of as a measure rather than as a density.

In fact, we already encountered in the previous lecture a diffusive variant of this model. The degree of simplification leading from the inelastic Boltzmann equation to (19) can probably be compared to that which leads from the compressible Navier-Stokes equation to Burgers' equation.

Let us consider the problem of asymptotic behavior for (19). Without loss of generality, we assume that the mean velocity is 0,

$$\int f_t(x)\, x\, dx = 0.$$

By elementary means it is easy to show

$$\frac{d}{dt} \int f_t(x)|x|^2 \, dx \leq -C \left(\int f_t(x)|x|^2 \, dx \right)^{3/2},$$

and from this one deduces

$$\sqrt{\int f_t(x)|x|^2 \, dx} = O\left(\frac{1}{t}\right)$$

as $t \to \infty$. It turns out that this rate is optimal. Note that

$$\sqrt{\int f_t(x)|x|^2 \, dx} = W_2(f_t\, dx, \delta_0).$$

In this oversimplified case, is there a self-similar solution ? After determining the scaling invariance of the equation, it is natural to set

$$\tau = \log t, \qquad g(\tau, x) = t f(t, tx) = e^\tau f(e^\tau, e^\tau x). \tag{20}$$

A few lines of computation lead to an equation on the new density g,

$$\frac{\partial g}{\partial \tau} = \frac{\partial}{\partial x} \left(g \frac{\partial}{\partial x} \left(g * W - \frac{x^2}{2} \right) \right). \tag{21}$$

So, again we recognize the same familiar structure as in the previous lecture,

$$\frac{\partial g}{\partial \tau} = \frac{\partial}{\partial x} \left[g \frac{\partial}{\partial x} \frac{\delta E}{\delta g} \right],$$

where now

$$E(g) = \frac{1}{2} \int g(x)g(y) \frac{|x-y|^3}{3} \, dx\, dy - \int g(x) \frac{|x|^2}{2} \, dx. \tag{22}$$

This functional E admits a whole family of critical points, but one unique minimizer, which is a singular measure,

$$g_\infty = \frac{1}{2} \left(\delta_{-\frac{1}{2}} + \delta_{\frac{1}{2}} \right).$$

A nontrivial theorem by Benedetto, Caglioti and Pulvirenti [6] states that if the initial datum g_0 is absolutely continuous with respect to Lebesgue measure, then g_τ converges (in weak measure sense) towards g_∞ as $\tau \to \infty$. In particular, for all $p \geq 1$ one has

$$W_p(g_\tau, g_\infty) \xrightarrow[\tau \to \infty]{} 0.$$

Note that in this context the use of Wasserstein distance is very natural, since the equilibrium is singular. It would *not* be possible to use L^1 distance, as in the previous lecture.

To g_∞ is associated a self-similar solution of (19), via (20). It reads

$$S_t = \frac{1}{2} \left(\delta_{-\frac{1}{2t}} + \delta_{\frac{1}{2t}} \right). \tag{23}$$

Moreover,

$$W_p(f_t, S_t) = \frac{W_p(g_\tau, g_\infty)}{t},$$

which shows that

$$W_p(f_t, S_t) = o \left(\frac{1}{t} \right). \tag{24}$$

Since on the other hand, $W(f_t, \delta_0) = O(1/t)$, we see that (23) indeed plays the role of a homogeneous cooling state.

Now can one get an estimate of how better is the approximation obtained by replacing δ_0 by S_t ? This amounts to study the *rate of convergence for the rescaled problem*. So a natural question is whether there is *exponential convergence to equilibrium* for solutions of (21). And in view of our study of the previous lecture, we could try a functional approach of this problem, by looking for a functional inequality of the form

$$\int g \left| \frac{\partial}{\partial x} \left(g * W - \frac{x^2}{2} \right) \right|^2 dx \geq \text{const.}[E(g) - E(g_\infty)].$$

It turns out that such an inequality is **false**, and this has to do with the negative sign in (22). If one thinks in terms of usual convexity, then the minus sign seems no worse than the positive sign, but if one thinks in terms of *displacement convexity*, then we see that the impact of this change is dramatic. Indeed, the functional $f \mapsto \int f(x)|x|^2 \, dx$ is uniformly displacement convex, while its negative is uniformly displacement concave.

In fact, one can prove that **there is no exponential convergence** at the level of (21). A recent result by Caglioti and the author establishes the estimate

$$\int_0^\tau W_p(g_s, g_\infty) \, ds \geq K \log \tau, \tag{25}$$

for some constant $K > 0$ depending on the initial datum. This inequality holds true as soon as g_0 (or f_0) is distinct from a convex combination of two symmetric dirac masses. As a consequence,

$$\int_0^T W_p(f_t, S_t) \, dt \geq K \log \log T,$$

which shows that "morally" $W_p(f_t, S_t)$ should not decrease to 0 faster than $1/(t \log t)$. This means that the improvement which one obtains when replacing δ_0 by S_t is at most logarithmic in time, hence very, very poor.

Again, here below is a precise theorem.

Theorem 3.1. *Let g_0 be a probability measure on $\mathbb{R}$, which is not a symmetric convex combination of two delta masses, and let g_τ be the corresponding solution to (21). Then, for all $p \in [1, +\infty)$,*

$$\int_0^{+\infty} W_p(g_\tau, g_\infty) \, d\tau = +\infty. \tag{26}$$

More precisely, there exists some constant $K > 0$, depending on g_0, such that, as $\tau \to \infty$,

$$\int_0^\tau W_p(g_s, g_\infty) \, ds \geq K \log \tau. \tag{27}$$

Corollary 3.1. *Let f_0 be a probability measure on $\mathbb{R}$, which is not a symmetric convex combination of two delta masses, and let f_t be the corresponding solution to (19). Then, for all $p \in [1, +\infty)$,*

$$\int_0^{+\infty} W_p(f_t, S_t) \, dt = +\infty. \tag{28}$$

More precisely, there exists some constant $K > 0$, depending on f_0, such that, as $T \to \infty$,

$$\int_0^T W_p(f_t, S_t) \, dt \geq K \log \log T. \tag{29}$$

In the rest of this lecture, I will try to convey an idea of how the proof works. One striking feature of it is that the introduction of the Wasserstein distance leads almost by itself to the solution. The key properties of the problem are (a) the presence of **vacuum** and **singularities** in the asymptotic state (one of these properties would suffice), (b) the structure of **nonlinear transport equation** in (21) (here we shall take advantage of the nonlinearity). The basic idea, stated informally, is that, due to the presence of vacuum and singularities, it requires a lot of work from the solution to approach the asymptotic state, which can be expressed by the strength of some velocity field; but the velocity field which drives the particles is coupled to the density, and approaches zero when the density approaches the asymptotic state. All in all, the convergence cannot be fast.

In the argument, let us assume for simplicity that f_0, and therefore g_0, is absolutely continuous with respect to Lebesgue measure (a property which is preserved by the flow). Rewrite equation (21) as

$$\frac{\partial g}{\partial \tau} + \frac{\partial}{\partial x}\left(g\,\xi[g]\right) = 0, \tag{30}$$

$$\xi[g](x) = x - \int g(y)(x-y)|x-y|\,dy. \tag{31}$$

Note that $\xi[g]$ lies in C^1 without any regularity assumption on g. Thus we can appeal to the usual theory of characteristics. If we introduce the vector fields T_τ solutions of

$$\frac{d}{d\tau}T_\tau(x) = \xi[g_\tau] \circ T_\tau(x),$$

we know from (30) that

$$g_\tau = T_\tau \# g_0. \tag{32}$$

Next, a remark which will simplify the analysis is that the ordering of particles is preserved. This is a general property of one-dimensional transport equations: two characteristic curves cannot cross because they are integral curves of a vector field. In particular, the **median** of the particles is preserved : if m stands for a median of the probability density g_0 at $\tau = 0$, then $m_\tau = T_\tau(m)$ is a median of g_τ. For simplicity, let us assume that there is just one median; then the density g_0 is strictly positive around its median m, in the sense that the interval $[m - \varepsilon, m + \varepsilon]$ carries a positive mass, for any $\varepsilon > 0$.

A fundamental quantity in a transport equation is the divergence of the velocity field. It provides a kind of measure on how fast the trajectories of the particles diverge. In dimension 1, this is just the derivative of the velocity field, and it is an easy exercice to deduce from (31) that $\xi[g]$ achieves its maximum on the set of medians of g (just differentiate (31) twice). In particular, $\xi[g_\tau]$ achieves its maximum precisely at m_τ.

We are now ready to explain the core of the proof of (25). To get a better intuition of what is going on, the reader is encouraged to draw pictures of particle trajectories. In order to converge towards the equilibrium state g_∞ as $\tau \to \infty$, it is necessary that particles which were initially infinitesimally close to each other, but located on different sides of the median, become asymptotically very far apart from each other (close to $-1/2$ or to $+1/2$).

This is possible only if the time-integral of the divergence of the velocity field diverges as $\tau \to \infty$. More precisely, if $a_- < m < a_+$ and $a_\pm(\tau) = T_t(a_\pm)$, then

$$\frac{d}{d\tau}[a_+(\tau) - a_-(\tau)] = \xi[g_\tau](a_+(\tau)) - \xi[g_\tau](a_-(\tau))$$

$$\leq \left\| \frac{d\xi[g_\tau]}{dx} \right\|_\infty [a_+(\tau) - a_-(\tau)],$$

so that

$$a_+(\tau) - a_-(\tau) \leq [a_+ - a_-] \exp \left(\int_0^\tau \left\| \frac{d\xi[g_s]}{dx} \right\|_\infty ds \right). \tag{33}$$

If we let $\tau \to \infty$, then from $a_-(\tau) \to -1/2$ and $a_+(\tau) \to +1/2$, we will deduce that

$$\exp \left(\int_0^\infty \left\| \frac{d\xi[g_s]}{dx} \right\|_\infty ds \right) \geq \frac{1}{[a_+ - a_-]},$$

and then by letting $a_+ - a_-$ go to 0, we find

$$\int_0^\infty \left\| \frac{d\xi[g_s]}{dx} \right\|_\infty ds = +\infty, \tag{34}$$

as announced.

But condition (34) will be extremely hard to achieve, because $\xi = 0$ at equilibrium. In fact, one has the continuity estimate, in Wasserstein distance,

$$\left\| \frac{d\xi_\tau}{dx} \right\|_\infty \leq 2W_1(g_\tau, g_\infty), \tag{35}$$

which implies that the divergence of ξ has to be all the smaller that one is close to equilibrium.

The proof of (35) is almost self-evident: on one hand, by direct computation,

$$\frac{d\xi_\tau}{dx} = 2 \int_{x \leq m_\tau} g_\tau(x) \left(x + \frac{1}{2} \right) dx - 2 \int_{x \geq m_\tau} g_\tau(x) \left(x - \frac{1}{2} \right) dx;$$

on the other hand,

$$W_1(g_\tau, g_\infty) = \int_{x \leq m_\tau} g(x) \left| x + \frac{1}{2} \right| dx + \int_{x \geq m_\tau} g(x) \left| x - \frac{1}{2} \right| dx.$$

The combination of (34) and (35) implies

$$\int_0^\infty W_1(g_\tau, g_\infty) ds = +\infty,$$

which is already an indication of slow decay to equilibrium. Under the assumption that g_0 is bounded below close to its median, it is possible to push the reasoning further in order to obtain the more precise result mentioned earlier. For this one just has to take advantage once more of the definition of the Monge-Kantorovich distance, and to establish that

$$W_1(g_\tau, g_\infty) \geq \frac{1}{2} \left(\int_{a_-(\tau)}^{a_+(\tau)} g_\tau \right) [1 - (a_+(\tau) - a_-(\tau))]$$

$$= \frac{1}{2} \left(\int_{a_-}^{a_+} g \right) [1 - (a_+(\tau) - a_-(\tau))].$$

Combining this with (33) and choosing $a_+ - a_-$ to be of the order of $W_1(g_\tau, g_\infty)$, one obtains a Gronwall-type inequality on this Wasserstein distance, which implies (25).

Note that here we have only taken advantage of the presence of vacuum in the support of the asymptotic state, not of the presence of singularities. Taking advantage of these singularities is slightly more complicated, but leads to (25) under more general assumptions, since one does not need to assume that the median is unique. The spirit is quite the same as before: if the support of the solution is made of more than just two points, this means that part of the particles will have distinct trajectories converging towards a single point, which is possible only if the time-integral of the divergence of the velocity field diverges to $-\infty$. Then this enters in conflict with the fact that the velocity field vanishes at equilibrium, and one can more or less repeat the same reasoning as above.

4 Estimates in a mean-field limit problem

In this lecture, I will explain about the use of some estimates from the theory of **concentration of measure** in some mean-field limit problems. The ideas below have been developed by a number of authors, in particular the group of Bakry and Ledoux in Toulouse. Since I will consider a stochastic microscopic model, I shall use a tiny bit of probabilistic formalism. The standard notations P and E will denote respectively the probability of an event, and the expectation of a measurable function. Whenever μ is a probability measure, F a measurable function and A a measurable set, I shall also use the convenient shorthand $\mu[F \in A] = \mu[\{x; F(x) \in A\}]$.

Recall the simple mean-field limit problem which was mentioned in the introduction: a random system of N particles, with respective positions $X_t^i \in \mathbb{R}^d$ ($1 \leq i \leq N$), obeying the system of stochastic differential equations

$$dX_t^i = dB_t^i - \nabla V(X_t^i) \, dt - \frac{1}{N} \sum_{j=1}^{N} \nabla W(X_t^i - X_t^j) \, dt. \tag{36}$$

The initial positions of the N particles are assumed to be random, independent and identically distributed. Moreover, the $(B_t^i)_{t\geq 0}^{1\leq i\leq N}$ are N independent Brownian motions in $\mathbb{R}^d$.

Under various assumptions on V, W, one can establish that the empirical measure at time t, which is a random measure, has a deterministic limit as $N \to \infty$:

$$\hat{\mu}_t^N = \frac{1}{N}\sum_{j=1}^N \delta_{X_t^i} \xrightarrow[N\to\infty]{} f_t(x)\,dx.$$

Here the convergence as $N \to \infty$ is in the sense of convergence in law of random variables, and for the weak topology of measures. Note that the left-hand side is a random measure, while the right-hand side is deterministic. The convergence can be restated as follows: let φ be a bounded smooth test-function, then, as $N \to \infty$,

$$E\left|\frac{1}{N}\sum_{j=1}^N \varphi(X_t^i) - \int_{\mathbb{R}^d} f_t(x)\,\varphi(x)\,dx\right| \xrightarrow[N\to\infty]{} 0.$$

The use of averages ("observables") like $(1/N)\sum \varphi(X_t^i)$ is quite natural. Moreover, the limit density f_t is characterized by its being a solution of the simple McKean-Vlasov equation

$$\frac{\partial f}{\partial t} = \frac{1}{2}\Delta f + \nabla \cdot (f\,\nabla V) + \nabla \cdot (f\,\nabla(f*W)). \tag{37}$$

As shown in Sznitman [35], this property of the empirical measure becoming deterministic in the limit is equivalent to the requirement that the law of the N particles be **chaotic** as $N \to \infty$, which means, roughly speaking, that k particles among N (k fixed, $N \to \infty$) look like independent, identically distributed random variables.

The problem addressed here is to find a quantitative version of this asymptotic behavior of the empirical measure. Let φ be a smooth test-function, can one find a bound on how much the average $(1/N)\sum \varphi(X_t^i)$ deviates from its asymptotic value $\int f_t(x)\varphi(x)\,dx$? More precisely, given some $\varepsilon > 0$, can one estimate

$$P\left(\left|\frac{1}{N}\sum_{i=1}^N \varphi(X_t^i) - \int_{\mathbb{R}^d} f_t(x)\varphi(x)\,dx\right| \geq \varepsilon\right) \quad ?$$

The motivations for this question do not only come from theoretical purposes, but also from numerical simulations and the will to justify the use of particle methods.

This problem is quite reminiscent of Sanov's theorem about large deviations for empirical measures. However, here the deviations are taken with respect to the limit value, so it has more to do with a law of large numbers, than with a large deviations principle. One possible way of attacking this question is the so-called theory of **concentration of measure**.

Let us give a very sketchy and elementary background about this theory, which is often traced back to works by Lévy, Gromov (linked with isoperimetric inequalities), and Milman. Roughly speaking, the main paradigm is as follows: let μ be some nice (to be precised later) reference measure on some measure space; for instance a gaussian measure on $\mathbb{R}^n$. Let A be some set of positive measure, say $\mu[A] \geq 1/2$. Then, in some sense, most points in our space are "not too far" from A. A typical example is that if μ is a gaussian measure on $\mathbb{R}^n$, and

$$A_t \equiv \{x \in \mathbb{R}^n; \, \text{dist}(x, A) \leq t\},$$

then $\mu[A_t] \to 1$ as $t \to \infty$, very quickly; more precisely, there exists $C, c > 0$ such that

$$\mu[A_t] \geq 1 - Ce^{-ct^2}.$$

The best possible rate c is obtained by the study of gaussian isoperimetry. It is important to note that the constants c and C will depend on a lower bound on the covariance matrix of the Gaussian, but not on the dimension n.

An equivalent way of formulating this principle consists in stating that "any Lipschitz function is not far from being constant". More precisely, if μ still denotes some gaussian measure, it is possible to show that whenever φ is a Lipschitz function on $\mathbb{R}^n$ with Lipschitz constant $\|\varphi\|_{\text{Lip}}$, then for all $r > 0$,

$$\mu\big[|\varphi - E_\mu \varphi| \geq r\big] \leq C \exp\left(-\frac{cr^2}{\|\varphi\|_{\text{Lip}}^2}\right). \tag{38}$$

Such principles are particularly useful in large dimension. It is easy to understand why: let φ be a Lipschitz function on $\mathbb{R}^d$, with Lipschitz constant $\|\varphi\|_{\text{Lip}}$, and let

$$\Phi(x^1, \ldots, x^N) = \frac{1}{N} \sum_{i=1}^N \varphi(x^i).$$

Then, an application of Cauchy-Schwarz inequality gives

$$\|\Phi\|_{\text{Lip}} \leq \frac{\|\varphi\|_{\text{Lip}}}{\sqrt{N}}. \tag{39}$$

Combining this with (38), we see that when N is large, Φ is sharply concentrated (with respect to μ) around its mean value.

During the last decade, the theory of concentration of measure was spectacularly developed by Talagrand in the framework of product spaces. On this occasion he introduced powerful and elementary induction methods to prove concentration inequalities in extremely general settings. Alternative, "functional" approaches have been developed by various authors, in particular Ledoux and Bobkov. One of their main contribution was to show how certain classes of functional inequalities enabled one to recover concentration inequalities, via more "global" and intrinsic methods (not depending so much on the space being product). All this is very well explained in the review paper by Ledoux [22] (in french) or in his lecture notes [23].

Typical classes of useful inequalities in this context are the logarithmic Sobolev inequalities, or the Poincaré inequalities. Let us recall their definitions. **Definition** *A probability measure μ satisfies a logarithmic Sobolev inequality with constant $\lambda > 0$ if for all $f \in L^1(d\mu)$ such that $f \geq 0$, $\int f\, d\mu = 1$, one has*

$$\int f \log f\, d\mu \leq \frac{1}{2\lambda} \int \frac{|\nabla f|^2}{f}\, d\mu.$$

It satisfies a Poincaré inequality with constant λ if for all $f \in L^2(d\mu)$,

$$\int f^2\, d\mu - \left(\int f\, d\mu \right)^2 \leq \frac{1}{\lambda} \int |\nabla f|^2\, d\mu.$$

It is known that if a probability measure satisfies a logarithmic Sobolev inequality with constant λ, then it also satisfies a Poincaré inequality with the same constant. Probability measures on $\mathbb{R}^n$ that admit a smooth, positive density behaving like $e^{-|x|^\alpha}$ as $|x| \to \infty$ satisfy a logarithmic Sobolev inequality if and only if $\alpha \geq 2$, and a Poincaré inequality if and only if $\alpha \geq 1$.

¿From the works of Ledoux, Bobkov, Götze and others it is known that if μ satisfies a logarithmic Sobolev inequality, then it satisfies concentration inequalities similar to those which are satisfied by a gaussian measure. Instead of explaining their methods, and in particular Ledoux's adaptation of the so-called "Herbst argument", I shall rather take a more indirect, but maybe more intuitive route towards this result. It will make the Wasserstein distance play a prominent role.

The following theorem is one of the main results of Otto and Villani [33] (the original version of the result contained a minor additional assumption, which was later removed by Bobkov et al. [7] with a very different proof).

Theorem 4.1. *Assume that μ has a smooth density and satisfies a logarithmic Sobolev inequality with constant $\lambda > 0$. Then it also satisfies the following inequality: for all $f \in L^1(d\mu)$, $f \geq 0$, $\int f\, d\mu = 1$,*

$$W_2(f\mu, \mu) \leq \sqrt{\frac{2}{\lambda} \int f \log f\, d\mu}. \tag{40}$$

In the particular case when μ is a gaussian measure, inequality (40) is due to Talagrand; accordingly we shall call (40) a Talagrand inequality.

Now, here is an idea of the original proof of theorem 4.1. Let V be defined by $d\mu(x) = e^{-V(x)} dx$. Introduce the auxiliary PDE

$$\frac{\partial f}{\partial s} = \Delta f - \nabla V \cdot \nabla f,$$

with initial datum $f(0, \cdot) = f$, and denote by $(f_s)_{s \geq 0}$ its solution. One can check that $f_s \longrightarrow 1$ as $s \to \infty$. In particular,

$$W_2(f\mu, f_0\mu) = 0; \qquad \int f_0 \log f_0 \, d\mu = \int f \log f \, d\mu, \qquad (41)$$

$$W_2(f\mu, f_s\mu) \xrightarrow[s \to \infty]{} W_2(f\mu, \mu); \qquad \int f_s \log f_s \, d\mu \xrightarrow[s \to \infty]{0} . \qquad (42)$$

It is shown in [33] that

$$\frac{d}{ds} W_2(f_s\mu, \mu) \leq \sqrt{\int \frac{|\nabla f|^2}{f} \, d\mu}. \qquad (43)$$

The function $t \mapsto W_2(f_s\mu, \mu)$ may not be differentiable, but then one can interpretate the derivative in (43) in distribution sense. On the other hand, it is easily checked that for each $s > 0$,

$$\frac{d}{ds} \sqrt{\int f_s \log f_s \, d\mu} = -\frac{\int \frac{|\nabla f_s|^2}{f_s} \, d\mu}{2\sqrt{\int f_s \log f_s \, d\mu}},$$

and by logarithmic Sobolev inequality we deduce that

$$\frac{d}{ds} \sqrt{\int f_s \log f_s \, d\mu} \leq -\sqrt{\frac{\lambda}{2}} \sqrt{\int \frac{|\nabla f_s|^2}{f_s} \, d\mu}. \qquad (44)$$

Finally, the combination of (43), (44) (both of them integrated from $s = 0$ to $+\infty$), (41) and (42) implies inequality (40).

Below are two important consequences of inequality (40), in the form of **concentration inequalities**. Both of them are valid in a very general framework (Polish spaces...).

Consequence 4.1.1 (Marton; Talagrand) *For all set A with positive measure, and for all $t \geq 0$,*

$$\mu[A_t] \geq 1 - \exp\left\{ -\frac{\lambda}{2} \left(t - \sqrt{\frac{2}{\lambda} \log \frac{1}{\mu[A]}} \right)^2 \right\}. \qquad (45)$$

Consequence 4.1.2 (Bobkov and Götze) *For all Lipschitz function φ, satisfying $\|\varphi\|_{\mathrm{Lip}} \leq 1$ and $\int \varphi \, d\mu = 0$, one has*

$$\forall t > 0 \qquad \int e^{t\varphi} \, d\mu \leq e^{\frac{t^2}{2\lambda}}. \tag{46}$$

Or equivalently, for all Lipschitz function φ,

$$\int e^{t\varphi} \, d\mu \leq \exp\left(t \int \varphi \, d\mu \right) \exp\left(\frac{t^2}{2\lambda} \|\varphi\|_{\mathrm{Lip}}^2 \right). \tag{47}$$

As a further consequence, for all Lipschitz function φ,

$$\mu\left[\varphi - \int \varphi \, d\mu \geq \varepsilon \right] \leq \exp\left(-\frac{\lambda \varepsilon^2}{2\|\varphi\|_{\mathrm{Lip}}^2} \right), \tag{48}$$

and of course, by symmetry,

$$\mu\left[\left| \varphi - \int \varphi \, d\mu \right| \geq \varepsilon \right] \leq 2 \exp\left(-\frac{\lambda \varepsilon^2}{2\|\varphi\|_{\mathrm{Lip}}^2} \right), \tag{49}$$

Proof (Proof of consequence 4.1.1). For any measurable set B with $\mu[B] > 0$, it is clear that $1_B \mu / \mu[B]$ is a probability measure. So whenever A and B are measurable subsets with positive measure, then

$$W_2\left(\frac{1_A}{\mu[A]}\mu, \frac{1_B}{\mu[B]}\mu \right) \leq W_2\left(\frac{1_A}{\mu[A]}\mu, \mu \right) + W_2\left(\mu, \frac{1_B}{\mu[B]}\mu \right).$$

Applying inequality (40) to both terms in the right-hand side, one finds

$$W_2\left(\frac{1_A}{\mu[A]}\mu, \frac{1_B}{\mu[B]}\mu \right) \leq \sqrt{\frac{2}{\lambda} \log \frac{1}{\mu[A]}} + \sqrt{\frac{2}{\lambda} \log \frac{1}{\mu[B]}}.$$

On the other hand, one can give the following interpretation of $W_2((1_A/\mu[A])\mu, (1_B/\mu[B])\mu))^2$. It is the minimal work necessary to transport all of the mass of the probability measure $(1_A/\mu[A])\mu$, onto the probability measure $(1_B/\mu[B])\mu$, taking into account that moving one unit of mass by a distance d costs d^2. Therefore, this is bounded by $d(A,B)^2$, with $d(A,B) = \inf\{d(x,y);\ x \in A,\ y \in B\}$. So in the end,

$$d(A,B) \leq \sqrt{\frac{2}{\lambda} \log \frac{1}{\mu[A]}} + \sqrt{\frac{2}{\lambda} \log \frac{1}{\mu[B]}}. \tag{50}$$

If we now choose $B = A_t^c$, we have $d(A,B) \geq t$ and $\mu[B] = 1 - \mu[A_t^c]$, and (50) transforms into (45) after appropriate rewriting.

Proof (Proof of consequence 4.1.2). Let μ satisfy (40), and let f be a non-negative L^1 function, such that $f\mu$ is a probability measure. First of all, by Hölder inequality,

$$W_1(f\mu, \mu) \leq W_2(f\mu, \mu).$$

A general duality principle for mass transportation, called the **Kantorovich duality**, and which in the case of W_1 is often called the **Kantorovich-Rubinstein theorem**, states that

$$W_1(f\mu, \mu) = \sup_{\|\varphi\|_{\mathrm{Lip}} \leq 1} \left(\int \varphi \, f d\mu - \int \varphi \, d\mu \right).$$

Thus, if φ satisfies the assumptions of the theorem ($\int \varphi \, d\mu = 0$, $\|\varphi\|_{\mathrm{Lip}} \leq 1$), one has

$$\int \varphi f \, d\mu \leq W_2(f\mu, \mu) \leq \sqrt{\frac{2}{\lambda} \int f \log f \, d\mu}$$

$$\leq \frac{t}{2\lambda} + \frac{1}{t} \int f \log f \, d\mu.$$

In particular,

$$\sup_f \left(\int \varphi f \, d\mu - \frac{1}{t} \int f \log f \, d\mu \right) \leq \frac{t}{2\lambda}, \tag{51}$$

where the supremum is taken over all $L^1(d\mu)$ functions f such that $f \, d\mu$ is a probability measure. ¿From standard Legendre duality, the supremum in (51) turns out to be

$$\frac{1}{t} \log \left(\int e^{t\varphi} \, d\mu \right).$$

This concludes the proof of (46). Then the equivalence between (46) and (47) is immediate. Finally, (48) follows by the choice $t = \lambda \varepsilon / \|\varphi\|_{\mathrm{Lip}}^2$.

Combining consequence 4.1.2 with (39), one obtains

Corollary 4.1. *Let μ satisfy a logarithmic Sobolev inequality with constant λ on $(\mathbb{R}^d)^N$. Then, whenever φ is a Lipschitz function with Lipschitz constant bounded by 1,*

$$\mu \left[\left| \frac{1}{N} \sum_{i=1}^{N} \varphi(x^i) - E_\mu \left(\frac{1}{N} \sum_{i=1}^{N} \varphi(x^i) \right) \right| \geq \varepsilon \right] \leq 2 \exp \left(-\frac{\lambda N \varepsilon^2}{2} \right). \tag{52}$$

In the end of this lecture, I describe how Malrieu [24, 25] has used the concentration inequalities above in the study of the mean-field model (36). The following result is taken from [24].

Theorem 4.2. *Consider the system* (36), *with initial positions distributed according to the chaotic distribution* $f_0^{\otimes N}$. *Assume that* V *is smooth and uniformly convex with constant* $\beta > 0$, *and that* W *is smooth, convex, even and has polynomial growth at infinity. Let* $(f_t)_{t \geq 0}$ *be the solution to the McKean-Vlasov equation* (37). *Further assume that* $\int f_0(x)|x|^p \, dx < +\infty$ *for some* p *large enough, that* $\int f_0 \log f_0 < +\infty$, *and that* f_0 *satisfies a logarithmic Sobolev inequality. Then, there exist various constants* $C > 0$ *such that*

(i) propagation of chaos holds uniformly in time: there exists a symmetric system of "particles" $(Y_t^1, \ldots, Y_t^N)$, *identically distributed and independent, with law* $f_t(x) \, dx$, *such that*

$$\sup_{t \geq 0} E|X_t^i - Y_t^i|^2 \leq \frac{C}{N};$$

(ii) For any Lipschitz function φ *with Lipschitz constant bounded by 1,*

$$P\left(\left|\frac{1}{N}\sum_{i=1}^N \varphi(X_t^i) - \int_{\mathbb{R}^d} \varphi(x) f_t(x) \, dx\right| \geq \sqrt{\frac{C}{N}} + r\right) \leq 2 \exp\left(-\frac{Nr^2}{2D_t}\right),$$

where D_t *is uniformly bounded as* $t \to \infty$;

(iii) let $\mu_t^{(1,N)}$ *be the law of* X_t^1. *Then, it converges exponentially fast to equilibrium as* $t \to \infty$, *uniformly in* N:

$$W_2(\mu_t^{(1,N)}, \mu_\infty) \leq C e^{-\lambda t}.$$

Remark 4.1. These results take important advantage of the uniform convexity of V. They do not hold in the case where, for instance, $V = 0$, even if W is uniformly convex. The problem comes from the fact that the quantity $\sum X_t^i$ is invariant under the action of the drift associated with the potential W, and undergoes only diffusion. As shown by Malrieu [25], one can fully remedy this problem by projecting the whole system onto the hyperplane ($\sum X^i = 0$), which amounts to getting rid of this "approximate conservation law".

Proof (Sketch of proof of Theorem 4.2). The first part is conceptually easy, but requires some skill in the computations. The idea to introduce the system Y_t^i, as a system of N independent particles, each of which has law f_t, was popularized by Sznitman [35]. It is very likely that X_t and Y_t are quite close, because, by Itô's formula, Y_t solves

$$dY_t^i = dB_t^i - \nabla V(Y_t^i) - \nabla(f_t * W)(Y_t^i).$$

With the help of this formula, a long calculation shows that, if

$$\alpha(t) = E[(X_t^1 - Y_t^1)^2],$$

then

$$\alpha'(t) \le -2\beta\alpha(t) + \frac{C}{\sqrt{N}}\alpha(t)^{1/2}.$$

Moreover, one is allowed to choose $Y_0^i = X_0^i$, so that $\alpha(0) = 0$. Then, Gronwall's lemma implies a bound like

$$\alpha(t)^{1/2} \le \frac{C}{\sqrt{N}}[1 - e^{-\beta t}]$$

(here, C stands for various positive constants).

In many problems of mean-field limit this first part is very important because it shows that the understanding of the simpler system (Y_t) is enough to get a good knowledge of the original system (X_t). Surprisingly, the sequel of the proof will take extremely little advantage of this fact.

If φ is a smooth Lipschitz test-function, from this first part we deduce immediately that

$$|E\varphi(X_t^1) - E\varphi(Y_t^1)| \le \sqrt{\frac{C}{N}}.$$

Since the particles are exchangeable, this also shows that

$$\left| E\left(\frac{1}{N}\sum_{i=1}^N \varphi(X_t^i)\right) - E\left(\frac{1}{N}\sum_{i=1}^N \varphi(Y_t^i)\right) \right| \le \sqrt{\frac{C}{N}}.$$

As a consequence,

$$P\left(\left| \frac{1}{N}\sum \varphi(X_t^i) - \int f_t(x)\varphi(x)\,dx \right| \ge \frac{C}{\sqrt{N}} + r\right)$$

$$= P\left(\left| \frac{1}{N}\sum \varphi(X_t^i) - E\left(\frac{1}{N}\sum_{i=1}^N \varphi(Y_t^i)\right) \right| \ge \frac{C}{\sqrt{N}} + r\right)$$

$$\le P\left(\left| \frac{1}{N}\sum \varphi(X_t^i) - E\left(\frac{1}{N}\sum_{i=1}^N \varphi(X_t^i)\right) \right| \ge r\right)$$

Now, to prove (ii) it is sufficient to show that the law μ of X_t satisfies a concentration inequality like (52); hence it is sufficient to prove that it satisfies a logarithmic Sobolev inequality. As a consequence of Itô's formula, this probability measure solves the diffusion equation

$$\frac{\partial\mu}{\partial t} = \frac{1}{2}\Delta\mu + \nabla\cdot(\nabla\mathbf{V}\mu), \tag{53}$$

where Δ and ∇ are differential operators acting on $L^1(\mathbb{R}^{dN})$, and

$$\mathbf{V}(x^1, \dots x^N) = \sum_{i=1}^{N} V(x^i) + \frac{1}{2N} \sum_{i,j} W(x^i - x^j).$$

One easily checks that $\mathbf{V}$ is uniformly convex with constant β. A nontrivial result by Bakry [2] (heavily relying on the linearity of the microscopic equation (53)) implies that μ_t satisfies a logarithmic inequality with constant $\lambda_t = [e^{-2\beta t}\lambda_0^{-1} + \beta^{-1}(1 - e^{-2\beta t})]^{-1}$, if μ_0 itself satisfies a logarithmic Sobolev inequality with constant λ_0. This ends the proof of (ii).

As for the proof of (iii), it is based on ideas quite similar to those of lecture 2 about the speed of approach to equilibrium. One of the key facts here is that both the function $\int f \log f \, d\mu$ and the Wasserstein distance behave very well as the dimension N increases. This allows one to provide an argument of trend to equilibrium which works "uniformly well with respect to the dimension." For more information, the reader can consult [24, 25].

These developments suggest many interesting problems. For instance, can the convexity assumptions on the interaction potential be relaxed ? Can phase transition phenomena occur if the potential is plainly non-convex ?

5 Otto's differential point of view

In this last lecture I shall present Otto's construction, which provides a re-interpretation of the optimal mass transportation problem for quadratic cost, in a point of view which is very appealing both from the point of view of fluid mechanics, and from the geometrical point of view. This re-interpretation is formal, and at first sight it is not clear what one gains by it; but it leads to an important harvest of results, some of which are explained in the end. If one keeps Otto's interpretation in mind, then some of the strange formulas encountered before in lectures 2 and 4, such as (11), (16) or (40), appear more natural. As a word of caution, let me insist that this interpretation is definitely not a universal explanation to all the interesting phenomena about mass transportation.

To begin with, let me present the **Benamou-Brenier formulation** of optimal transportation with quadratic cost.

Let ρ_0 and ρ_1 be two probability density on $\mathbb{R}^n$, say with compact support. Think of them as describing two different states of a certain bunch of particles. Assume now that at time $t = 0$, this bunch of particles is in the state described by ρ_0, and you have the possibility to make them move around by imposing any time-dependent velocity field you wish in $\mathbb{R}^n$. This has an energetical cost, which coincides with the total kinetic energy of the particles. Your goal is to have, at time $t = 1$, the particles in state ρ_1, and furthermore you are looking for the solution which will require the least amount of work. Stated informally, your goal is to minimize the **action**

$$A = \int_0^1 \left(\sum_x |\dot{x}(t)|^2 \right) dt,$$

with x varying in the set of all particles, and x at time 0 is distributed according to ρ_0, at time 1 according to ρ_1.

This can be rephrased in a purely Eulerian way: let ρ_t be the density of the bunch of particles at time t, and v_t the associated velocity field (defined only on the support of ρ_t). Then the preceding problem becomes

$$\inf_{\rho,v} \left\{ \int_0^1 \int \rho_t(x) |v_t(x)|^2 \, dx \, dt; \quad \frac{\partial \rho_t}{\partial t} + \nabla \cdot (\rho_t v_t) = 0 \right\}, \qquad (54)$$

where the infimum is taken over all time-dependent probability densities $(\rho_t)_{0 \le t \le 1}$ which agree with ρ_0 and ρ_1 at respective times $t = 0$ and $t = 1$, and overall time-dependent velocity fields $(v_t)_{0 \le t \le 1}$ which convect ρ_t, as expressed by the continuity equation in the right-hand side of (54).

This minimization problem is not very precisely defined as far as functional spaces are concerned. If one wants to give a mathematically rigorous definition, it is best to change unknowns (ρ, v) for $(\rho, P) = (\rho, \rho v)$ and to rewrite (54) as

$$\inf_{\rho, P} \int_0^1 \int \frac{|P_t(x)|^2}{\rho_t(x)} \, dx \, dt; \quad \frac{\partial \rho_t}{\partial t} + \nabla \cdot P_t = 0. \qquad (55)$$

Then this minimization problem makes sense under minimal regularity assumptions on (ρ, P) (say continuous with respect to time, ρ taking values in probability space and P in vector-valued distributions). This comes from the joint convexity of $|P|^2/\rho$ as a function of P and ρ, and from the linearity of constraints.

Theorem 5.1 (Benamou-Brenier). *Assume that ρ_0, ρ_1 are probability densities on $\mathbb{R}^n$ with compact support, and let I be the value of the infimum in (55). Then*

$$I = \min_{T\#\rho_0 = \rho_1} \int |x - T(x)|^2 \, \rho_0(x) \, dx = W_2(\rho_0, \rho_1)^2. \qquad (56)$$

Note that in this statement, we have again identified measures with their densities. The second inequality in (56) is part of the Brenier-McCann theorem alluded to in lecture 2, so the new point is only that this infimum also coincides with I.

Proof (Sketch of proof). Here is a formal proof of (56), in which regularity issues are discarded. Let us first check that

$$I \ge W_2(\rho_0, \rho_1)^2.$$

Let (ρ, v) be admissible in (55), and introduce the characteristics associated with v. These are the integral curves $(T_t x)_{0 \leq t \leq 1}$ defined by

$$\frac{d}{dt} T_t x = v_t(T_t x).$$

¿From the theory of linear transport equations it is known that $\rho_t = T_t \# \rho_0$; in particular $\rho_1 = T_1 \# \rho_0$. Then,

$$\int \rho_t(x) |v_t(x)|^2 \, dx = \int \rho_0(x) |v_t(T_t x)|^2 \, dx$$

$$= \int \rho_0(x) \left| \frac{d}{dt} T_t x \right|^2 \, dx.$$

Therefore, integrating in t and using Jensen's inequality, we find

$$\int_0^1 \int \rho_t(x) |v_t(x)|^2 \, dx \, dt \geq \int \rho_0(x) \left| \int_0^1 \left(\frac{d}{dt} T_t x \right) dt \right|^2 dx$$

$$= \int \rho_0(x) |T_1(x) - T_0(x)|^2 \, dx$$

$$= \int \rho_0(x) |T_1(x) - x|^2 \, dx.$$

Since $T_1 \# \rho_0 = \rho_1$, this quantity is bounded below by the right-hand side of (56).

In a second step, let us check the reverse inequality

$$I \leq W_2(\rho_0, \rho_1)^2.$$

Let $T = \nabla \varphi$ be given by the Brenier-McCann theorem: it satisfies $T \# \rho_0 = \rho_1$ and

$$W_2(\rho_0, \rho_1)^2 = \int \rho_0(x) |x - T(x)|^2 \, dx.$$

¿From this T it is easy to construct an admissible (ρ, v): just set

$$T_t(x) = (1 - t)x + tT(x), \qquad v_t = \left(\frac{d}{dt} T_t \right) \circ T_t^{-1},$$

$$\rho_t = T_t \# \rho_0.$$

Then one easily checks that for all $t \in [0, 1]$,

$$\int \rho_t(x) |v_t(x)|^2 \, dx = \int \rho_0(x) |T(x) - x|^2 \, dx = W_2(\rho_0, \rho_1)^2.$$

In particular,

$$\int_0^1 \left(\int \rho_t |v_t|^2 \right) dt = W_2(\rho_0, \rho_1)^2,$$

which implies the conclusion.

It was noticed by Otto that theorem 5.1 leads to the re-interpretation of the Wasserstein distance as the geodesic distance for some "Riemannian" structure on the set of probability densities:

$$W_2(\rho_0, \rho_1)^2 = \inf \left\{ \int_0^1 \left\| \frac{\partial \rho}{\partial t} \right\|_{\rho(t)}^2 \, dt; \quad \rho(0) = \rho_0, \ \rho(1) = \rho_1, \right\}$$

where the "metric" on the "tangent space" at ρ would be defined by

$$\left\| \frac{\partial \rho}{\partial s} \right\|_\rho^2 = \inf_v \left\{ \int \rho |v|^2; \quad \frac{\partial \rho}{\partial s} + \nabla \cdot (\rho v) = 0 \right\}.$$

Here is the interpretation: let $\partial \rho / \partial s$ be an infinitesimal variation of the probability density ρ. If one thinks of ρ as the density of a bunch of particles, this infinitesimal variation can be associated to many possible velocity fields for the particles: namely, all those fields v's which satisfy the continuity equation $\nabla \cdot (\rho v) = -\partial \rho / \partial s$. To each of these velocity fields is associated a kinetic energy, which is the total kinetic energy of the particles. Now, to this infinitesimal variation associate the minimum of the kinetic energy, for all possible choices of the velocity field.

Assume that ρ is smooth and positive, then an optimal v should be characterized by its being a *gradient*. Indeed, let v be optimal, and let w be any divergence-free vector field, then $v_\varepsilon = v + \varepsilon(w/\rho)$ is another admissible vector field, so the associated kinetic energy is no less than the kinetic energy associated to v. Thus,

$$\int \rho |v|^2 \leq \int \rho |v_\varepsilon|^2 = \int \rho |v|^2 + 2\varepsilon \int v \cdot w + \varepsilon^2 \int \frac{|w|^2}{\rho}.$$

Simplifying and letting $\varepsilon \to 0$, we see that $\int v \cdot w = 0$. Since w was arbitrary in the space of divergence-free vector fields, v should be a gradient.

Note that, at least formally, the elliptic PDE

$$-\nabla \cdot (\rho \nabla \varphi) = \frac{\partial \rho}{\partial s} \tag{57}$$

should be uniquely solvable (up to an additive constant) for φ.

Remark 5.1. If we wished to discuss the whole thing in a bounded subset of $\mathbb{R}^n$, rather than in the whole of $\mathbb{R}^n$, then we should choose a convex set containing the supports of ρ_0 and ρ_1, and impose Neumann boundary condition for (57).

To sum up, in Otto's formalism the Wasserstein distance is the geodesic distance associated to the following metric: for each probability density ρ, whenever $\partial \rho / \partial s$ and $\partial \rho / \partial t$ are any two infinitesimal variations of ρ, define

$$\left\langle \frac{\partial \rho}{\partial s}, \frac{\partial \rho}{\partial t} \right\rangle_\rho = \int \rho \nabla \varphi \cdot \nabla \psi, \tag{58}$$

where φ, ψ are the solutions of

$$-\nabla \cdot (\rho \nabla \varphi) = \frac{\partial \rho}{\partial s}, \qquad -\nabla \cdot (\rho \nabla \psi) = \frac{\partial \rho}{\partial t}.$$

The construction above cannot apparently be put on solid ground, in the sense that it seems hopeless to properly define a Riemannian structure along the lines suggested above. However, this formalism has two main advantages: first, in some situations it backs up intuition in a powerful way; second, it suggests very efficient rules of formal computations. Indeed, as soon as one has defined a Riemannian structure, then associated to it are rules of **differential calculus**: in particular, one can define a **gradient** and a **Hessian** operators. Recall the general definition of a gradient operator: it is defined by the identity

$$\left\langle \operatorname{grad} F(\rho), \frac{\partial \rho}{\partial s} \right\rangle_\rho = DF(\rho) \cdot \frac{\partial \rho}{\partial s},$$

where F is any ("smooth") function on the manifold into consideration, and DF the differential of F. With the above definitions, it can be checked that the gradient of a function F on the set of probability densities is defined by

$$\operatorname{grad} F(\rho) = -\nabla \cdot \left(\rho \nabla \frac{\delta F}{\delta \rho} \right), \tag{59}$$

where $\delta F/\delta \rho$ denotes, as in lecture 2, the gradient of the functional F with respect to the L^2 Euclidean structure.

As for Hessian operators, it is not very interesting to give a general formula, but one can often reconstruct them from the formula

$$\left\langle \operatorname{Hess} F(\rho) \cdot \frac{\partial \rho}{\partial s}, \frac{\partial \rho}{\partial s} \right\rangle = \frac{d^2}{ds^2} \bigg|_{s=0} F(\rho_s),$$

with $(\rho_s)_{s \geq 0}$ standing for the geodesic curve issued from ρ with velocity $\partial \rho/\partial s$. This geodesic can be described as follows: let φ be associated to $\partial \rho/\partial s$ via (57), then

$$\rho_s = [(1 - s)\mathrm{id} + s\varphi] \# \rho.$$

As a consequence of (59), equations with the particular structure (5) can be seen as **gradient flows** of the energy F with respect to Otto's differential structure. This is very interesting if one looks at the long-time or the short-time behavior of equations such as (5). Indeed, there are strong connections between the behavior of a gradient flow,

$$\frac{\partial f}{\partial t} = -\operatorname{grad} F(\rho)$$

and the convexity properties of F. In particular, here we understand that the natural notion of convexity, for the asymptotic properties of (5), is the convexity in the sense of Otto's differential calculus, i.e. convexity along geodesics, which coincides with the **displacement convexity** discussed in section 2. We also understand that formula (9) is just the natural Taylor formula in this context, to quantify how convex the energy is.

Here is an example of what can be deduced from these considerations.

Formal theorem to be checked on each case [Otto and Villani] *Let F be a λ-uniformly displacement convex functional. Then it admits a unique minimizer ρ_∞, and one has the functional inequality*

$$\int \rho \left| \nabla \frac{\delta F}{\delta \rho} \right|^2 \geq 2\lambda[F(\rho) - F(\rho_\infty)]. \tag{60}$$

As an example, let

$$F(\rho) = \frac{1}{m-1} \int \rho(x)^m \, dx + \int \rho(x) \frac{|x|^2}{2} \, dx$$

be defined on $L^1(\mathbb{R}^n)$. ¿From a theorem of McCann [27], the first half of the right-hand side is displacement convex if $m \geq 1 - 1/n$. On the other hand, the second half is 1-uniformly convex. Moreover, if $m \geq 1 - 1/n$, then the functional F is bounded below on the set of probability densities, with the exception of the case $m = 1/2$, $n = 1/2$. In all the other cases it follows from the formal theorem above that

$$\int \rho \left| \frac{m}{m-1} [\nabla \rho^{m-1}](x) + x \right|^2 \geq 2[F(\rho) - F(\rho_\infty)]. \tag{61}$$

These inequalities are the key to the study of the asymptotic behavior to the solutions of the porous medium equations,

$$\frac{\partial \rho}{\partial t} = \Delta \rho^m,$$

and were found independently by several authors [13, 18, 32].

The following astonishing remark by Dolbeault and Del Pino establishes an unexpected connection between this field and the theory of optimal Sobolev-type inequalities. Expand the square in the left-hand side of (61) and notice that the terms in $\int \rho(x)|x|^2 \, dx$ cancel out on both sides of the inequality. Then one is left with

$$\left(\frac{m}{m-1} \right)^2 \int \rho |\nabla \rho^{m-1}|^2 + \frac{2m}{m-1} \int \rho(x) \, (\nabla \rho^{m-1})(x) \cdot x \geq \frac{2}{m-1} \int \rho^m + C_\infty,$$

for some constant $C_\infty = -2F(\rho_\infty)$. It is not difficult to identify ρ_∞:

$$\rho_\infty(x) = \left(\sigma^2 + \frac{1-m}{2m} |x|^2 \right)_+^{\frac{1}{m-1}}$$

(in physics, this profile is known as the Barenblatt profile), and therefore one can compute the constant C_∞ explicitly. Next, by chain-rule and integration by parts,

$$\frac{2m}{m-1} \int \rho(x)\,(\nabla \rho^{m-1})(x) \cdot x = -2 \int \rho^m.$$

Moreover, by chain-rule,

$$\left(\frac{m}{m-1}\right)^2 \int \rho |\nabla \rho^{m-1}|^2 = \left(\frac{2m}{2m-1}\right)^2 \int |\nabla(\rho^{m-\frac{1}{2}})|^2.$$

On the whole, we see that

$$\left(\frac{2m}{2m-1}\right)^2 \int |\nabla(\rho^{m-\frac{1}{2}})|^2 \geq \frac{2m}{m-1} \int \rho^m + C_\infty,$$

for any probability density ρ. Or equivalently, for any nonnegative function ρ,

$$\left(\frac{2m}{2m-1}\right)^2 \left(\int \rho\right)^{1-m} \int |\nabla(\rho^{m-\frac{1}{2}})|^2 \geq \frac{2m}{m-1} \left(\int \rho^m\right) + C_\infty \left(\int \rho\right)^m.$$

If one now sets $u = \rho^{m-1/2}$, this transforms into

$$\left(\frac{2m}{2m-1}\right)^2 \|\nabla u\|_{L^2} \|u\|_{L^{2/(2m-1)}}^{1-m} \geq \frac{2m}{m-1} \|u\|_{L^{2m/(2m-1)}}^m + C_\infty \|u\|_{L^{2/(2m-1)}}^m.$$

By tedious but easy homogeneity argument, this transforms into a Gagliardo-Nirenberg inequality

$$\|u\|_{L^p} \leq C \|\nabla u\|_{L^2}^\theta \|u\|_{L^q}^{1-\theta}, \tag{62}$$

for some exponents p, q defined as follows: either $2 < p \leq 2n/(n-2)$ and $q = (p+2)/2$, or $0 < p < 2$ and $q = 2(p-1)$. Once p and q are given, the value of θ is completely determined by scaling homogeneity.

It turns out that the constants C obtained in this process are optimal ! and that the minimizers in these inequalities are obtained by a simple rescaling of the Barenblatt profile. Further note that the case $p = 2n/(n-2)$ corresponds to $\theta = 1$, and then (62) is nothing but the usual Sobolev embedding with optimal constants (for $n \geq 3$); it also corresponds to the critical exponent $m = 1 - 1/n$ found by McCann [27].

To conclude this lecture, let me mention an important application of Otto's interpretation: it suggests a natural approximation of gradient flow equations like (5) by a discrete time-step minimization procedure, see in particular [21]. This procedure can be extended to more general cost functions (in the definition of the mass transportation problem) than the quadratic one, see Otto [30], or Agueh's PhD thesis. For instance, the heat equation, but also the porous medium equations, the p-Laplace equations (or rather p-heat equations), or even a combination of both, can be obtained in this way. It can also be used for some Hamiltonian equations, see Cullen and Maroofi [17]. Otto has also worked on more degenerate examples [31] arising from fluid mechanics problems. Finally, Carlen and Gangbo have studied interesting variants of this time-step procedure in a kinetic context [10, 11].

References

1. ANÉ, C., BLACHÈRE, S., CHAFAÏ, D., FOUGÈRES, P., GENTIL, I., MALRIEU, F., ROBERTO, C., AND SCHEFFER, G. *Sur les inégalités de Sobolev logarithmiques*, vol. 10 of *Panoramas et Synthèses*. Société Mathématique de France, 2000.
2. BAKRY, D. On Sobolev and logarithmic Sobolev inequalities for Markov semigroups. In *New trends in stochastic analysis (Charingworth, 1994), Taniguchi symposium* (1997), World Sci. Publishing.
3. BAKRY, D., AND EMERY, M. Diffusions hypercontractives. In *Sém. Proba. XIX*, no. 1123 in Lecture Notes in Math. Springer, 1985, pp. 177–206.
4. BARTHE, F. On a reverse form of the Brascamp-Lieb inequality. *Invent. Math. 134*, 2 (1998), 335–361.
5. BENEDETTO, D., CAGLIOTI, E., CARRILLO, J. A., AND PULVIRENTI, M. A non-Maxwellian steady distribution for one-dimensional granular media. *J. Statist. Phys. 91*, 5-6 (1998), 979–990.
6. BENEDETTO, D., CAGLIOTI, E., AND PULVIRENTI, M. A kinetic equation for granular media. *RAIRO Modél. Math. Anal. Numér. 31*, 5 (1997), 615–641. Erratum in *M2AN Math. Model. Numer. Anal. 33*, 2 (1999), 439–441.
7. BOBKOV, S., GENTIL, I., AND LEDOUX, M. Hypercontractivity of Hamilton-Jacobi equations. *J. Math. Pures Appl. 80*, 7 (2001), 669–696.
8. BOBYLEV, A. V., CARRILLO, J., AND GAMBA, I. On some properties of kinetic and hydrodynamic equations for inelastic interactions. Preprint.
9. BRENIER, Y. Polar factorization and monotone rearrangement of vector-valued functions. *Comm. Pure Appl. Math. 44*, 4 (1991), 375–417.
10. CARLEN, E., AND GANGBO, W. Constrained steepest descent in the 2-Wasserstein metric. Preprint, 2001.
11. CARLEN, E., AND GANGBO, W. On the solution of a model Boltzmann equation via steepest descent in the 2-Wasserstein metric. Preprint, 2001.
12. CARRILLO, J., McCANN, R., AND VILLANI, C. Kinetic equilibration rates for granular media. Preprint, 2001.
13. CARRILLO, J. A., AND TOSCANI, G. Asymptotic L^1-decay of solutions of the porous medium equation to self-similarity. *Indiana Univ. Math. J. 49*, 1 (2000), 113–142.
14. CERCIGNANI, C. Recent developments in the mechanics of granular materials. In *Fisica matematica e ingeneria delle strutture : rapporti e compatibilità*, G. Ferrarese, Ed. Pitagora Ed., Bologna, 1995, pp. 119–132.
15. CORDERO-ERAUSQUIN, D. Some applications of mass transport to Gaussian type inequalities. To appear in *Arch. Rational Mech. Anal.*
16. CORDERO-ERAUSQUIN, D., GANGBO, W., AND HOUDRÉ, C. Inequalities for generalized entropy and optimal transportation. Preprint, 2001.
17. CULLEN, M., AND MAROOFI, H. The fully compressible semi-geostrophic system from meteorology. Preprint, 2001.
18. DEL PINO, M., AND DOLBEAULT, J. Generalized Sobolev inequalities and asymptotic behaviour in fast diffusion and porous medium problems. Preprint Univ. Paris IX-Dauphine, CEREMADE, n. 9905, 1999.
19. DEL PINO, M., AND DOLBEAULT, J. Best constants for Gagliardo-Nirenberg inequalities and applications to nonlinear diffusions. To appear in *J. Funct. Anal.*, 2001.

20. GROSS, L. Logarithmic Sobolev inequalities and contractivity properties of semigroups. In *Dirichlet forms (Varenna, 1992)*. Springer, Berlin, 1993, pp. 54–88. Lecture Notes in Math., 1563.

21. JORDAN, R., KINDERLEHRER, D., AND OTTO, F. The variational formulation of the Fokker-Planck equation. *SIAM J. Math. Anal. 29*, 1 (1998), 1–17.

22. LEDOUX, M. Inégalités isopérimétriques en analyse et probabilités. *Astérisque*, 216 (1993), Exp. No. 773, 5, 343–375. Séminaire Bourbaki, Vol. 1992/93.

23. LEDOUX, M. Concentration of measure and logarithmic Sobolev inequalities. In *Séminaire de Probabilités, XXXIII*. Springer, Berlin, 1999, pp. 120–216.

24. MALRIEU, F. Logarithmic Sobolev inequalities for some nonlinear PDE's. To appear in *Stochastic Process. Appl.*

25. MALRIEU, F. Convergence to equilibrium for granular media equations and their Euler schemes. Preprint, 2001.

26. MCCANN, R. J. Existence and uniqueness of monotone measure-preserving maps. *Duke Math. J. 80*, 2 (1995), 309–323.

27. MCCANN, R. J. A convexity principle for interacting gases. *Adv. Math. 128*, 1 (1997), 153–179.

28. MCNAMARA, S., AND YOUNG, W. Kinetics of a one-dimensional granular medium. *Phys. Fluids A 5*, 1 (1993), 34–45.

29. MURATA, H., AND TANAKA, H. An inequality for certain functional of multidimensional probability distributions. *Hiroshima Math. J. 4* (1974), 75–81.

30. OTTO, F. Double degenerate diffusion equations as steepest descent. Preprint Univ. Bonn, 1996.

31. OTTO, F. Lubrication approximation with prescribed nonzero contact angle. *Comm. Partial Differential Equations 23*, 11-12 (1998), 2077–2164.

32. OTTO, F. The geometry of dissipative evolution equations: the porous medium equation. *Comm. Partial Differential Equations 26*, 1-2 (2001), 101–174.

33. OTTO, F., AND VILLANI, C. Generalization of an inequality by Talagrand and links with the logarithmic Sobolev inequality. *J. Funct. Anal. 173*, 2 (2000), 361–400.

34. OTTO, F., AND VILLANI, C. Comment on: "Hypercontractivity of Hamilton-Jacobi equations" [J. Math. Pures Appl. (9) 80 (2001), no. 7, 669–696; 1846020] by S. G. Bobkov, I. Gentil and M. Ledoux. *J. Math. Pures Appl. (9) 80*, 7 (2001), 697–700.

35. SZNITMAN, A.-S. Topics in propagation of chaos. In *École d'Été de Probabilités de Saint-Flour XIX—1989*. Springer, Berlin, 1991, pp. 165–251.

36. TALAGRAND, M. Transportation cost for Gaussian and other product measures. *Geom. Funct. Anal. 6*, 3 (1996), 587–600.

37. TANAKA, H. An inequality for a functional of probability distributions and its application to Kac's one-dimensional model of a Maxwellian gas. *Z. Wahrscheinlichkeitstheorie und Verw. Gebiete 27* (1973), 47–52.

38. TANAKA, H. Probabilistic treatment of the Boltzmann equation of Maxwellian molecules. *Z. Wahrsch. Verw. Gebiete 46*, 1 (1978/79), 67–105.

39. VILLANI, C. A survey of mathematical topics in kinetic theory. To appear in *Handbook of fluid mechanics*, S. Friedlander and D. Serre, Eds.

40. VILLANI, C. Mass transportation, transportation inequalities and dissipative equations. Notes for a summer school in Ponta Delgada, September 2000, 2001.

41. VILLANI, C. Topics in optimal transportation. Forthcoming book by the AMS.

Extended Monge-Kantorovich Theory

Yann Brenier

CNRS, LJAD, Université de Nice,
en détachement de l'Université Paris 6, France,
brenier@math3.unice.fr

1 Abstract

We extend in various ways the Monge-Kantorovich theory (MKT), also known as optimal transportation theory (OTT) [Ka], [KS], [36], [Su]. This theory has become familiar in the last ten years in the field of nonlinear PDEs, especially because of its connection with the Monge-Ampère equation [Br0], [Br1], [Ca], [CP], [GM], the Eikonal equation [EGn], and the heat equation [JKO], [Ot1], [Ot2]... The first and crucial step of *all* our extensions consists in revisiting the MKT as a theory of generalized geodesics, following [BB]. Then, various generalizations of the MKT can be investigated, including a relativistic heat equations and a variational interpretation of Moser's lemma. Next, we define generalized harmonic functions and open several questions. Then, we consider multiphase MKT with constraints, which includes the relaxed theory of geodesics on groups of volume preserving maps related to incompressible fluid Mechanics [Br3]. Finally, we consider generalized extremal surfaces and we relate them to classical Electrodynamics, namely to the Maxwell equations and to the pressureless Euler-Maxwell equations.

2 Generalized geodesics and the Monge-Kantorovich theory

2.1 Generalized geodesics

Although we could consider the general framework of a Riemannian manifold, we only address the case of a subset D of the Euclidean space $\mathbf{R}^d$, and we assume D to be the closure of a convex open bounded set. Given two points X_0 and X_1 in D, the geodesic curve

$$X(s) = (1 - s)X_0 + sX_1 \tag{1}$$

achieves

$$\inf_X \int_0^1 k(X'(s))ds, \tag{2}$$

for all continuous convex even function k on $\mathbf{R}^d$, among all smooth paths $s \in [0,1] \to X(s) \in D$ such that $X(1) = X_1$, $X(0) = X_0$. This immediately follows from Jensen's inequality. In the spirit of Young's generalized functions [Yo], [Ta], let us now associate to each admissible path X the following pair of (Borel) measures (ρ, E) defined on the compact set $[0,1] \times D$ by

$$\rho(s,x) = \delta(x - X(s)), \quad E(s,x) = X'(s)\delta(x - X(s)), \quad (s,x) \in [0,1] \times D. \tag{3}$$

They satisfy the following compatibility condition in the sense of distributions

$$\partial_s \rho + \nabla \cdot E = 0, \quad \rho(0, \cdot) = \rho_0, \quad \rho(1, \cdot) = \rho_1, \tag{4}$$

where

$$\rho_0(x) = \delta(x - X_0), \quad \rho_1(x) = \delta(x - X_1). \tag{5}$$

Indeed, we have

$$-\int_D \int_0^1 (\partial_s \phi(s,x)d\rho(s,x) + \nabla \phi(s,x) \cdot dE(s,x)) \tag{6}$$

$$+ \int_D (\phi(1,x)d\rho_1(x) - \phi(0,x)d\rho_0(x)) = 0,$$

for all smooth functions $\phi(s,x)$ defined on $[0,1] \times \mathbf{R}^d$. (This also implies, in a weak sense, that E is parallel to the boundary ∂D.) We notice that E is absolutely continuous with respect to ρ and, by Jensen's inequality,

$$\int_0^1 k(X'(s))ds \tag{7}$$

is bounded from below by

$$K(\rho, E) = \int k(e)d\rho, \tag{8}$$

where $e(s,x)$ is the Radon-Nikodym derivative of E with respect to ρ. A more precise definition of K can be given in terms of the Legendre-Fenchel transform of k denoted by k^* and defined by

$$k^*(y) = \sup_{x \in \mathbb{R}^d} x \cdot y - k(x), \tag{9}$$

where $\cdot$ denotes the inner product in $\mathbb{R}^d$. We assume k^* to be continuous on $\mathbb{R}^d$. Typically

$$k(x) = \frac{|x|^p}{p}, \quad k^*(y) = \frac{|y|^q}{q}, \quad \frac{1}{p} + \frac{1}{q} = 1, \quad 1 < p, q < +\infty,$$

where $|\cdot|$ denotes the Euclidean norm in $\mathbb{R}^d$. We have

$$K(\rho, E) = \sup_{\alpha, \beta} \int_D \int_0^1 \alpha(s, x) d\rho(s, x) + \beta(s, x) \cdot dE(s, x), \tag{10}$$

where the supremum is performed over all pair (α, β) of respectively real and vector valued continuous defined on $[0, 1] \times D$ subject to satisfy

$$\alpha(s, x) + k^*(\beta(s, x)) \leq 0 \tag{11}$$

pointwise. (Indeed, it can be easily checked that, with this definition, $K(\rho, E)$ is infinite unless i) ρ is nonnegative, ii) E absolutely continuous with respect to ρ and has a Radon-Nikodym density e, ii) $K(\rho, E)$ is just the ρ integral of $k(e)$.) Notice that K is a convex functional, valued in $[0, +\infty]$.

It is now natural to consider the infimum, denoted by $\inf K$, of functional K defined by (10) among *all* pairs (ρ, E) that satisfy compatibility conditions (4), with data (5), and *not only* among those which are of form (3). This new minimization problem is convex (as the original one). Since the class of admissible solutions has been enlarged, the following upper bound follows

$$\inf K \leq k(X_1 - X_0) \tag{12}$$

(by using (1) and (3) as an admissible pair). It turns out that there is no gap between the original infimum and the relaxed one.

Theorem 2.1. *The infimum of functional K, defined by (10), among all pair $(\rho, E)(s, x)$ of measures on $[0, 1] \times D$, satisfying (4), with boundary conditions*

$$\rho(0, x) = \delta(x - X_0), \quad \rho(1, x) = \delta(x - X_1), \tag{13}$$

is achieved by the one associated, through (3), to the straight path between the end points $X(s) = (1 - s)X_0 + sX_1$.

Proof

The proof is obtained through the following simple, and typical, duality argument that will be used several times subsequently in these lecture notes. First, we use (6) to relax constraint (4) and write

$$\inf K = \inf_{\rho, E} \sup_{\alpha, \beta, \phi} \int_D \int_0^1 (\alpha(s, x) - \partial_s \phi(s, x)) d\rho(s, x) \qquad (14)$$

$$+(\beta(s, x) - \nabla\phi(s, x)) \cdot dE(s, x) + \int_D (\phi(1, x) d\rho_1(x) - \phi(0, x) d\rho_0(x)),$$

where (α, β) are subject to (11), and ϕ should be considered as a Lagrange multiplier for condition (4).

The *formal* optimality conditions for (α, β, ϕ) are

$$\alpha = \partial_s \phi, \quad \beta = \nabla\phi, \quad \alpha + k^*(\beta) = 0, \qquad (15)$$

which leads to the Hamilton-Jacobi equation

$$\partial_s \phi + k^*(\nabla\phi) = 0. \qquad (16)$$

Thus, a good guess for (α, β, ϕ) is

$$\phi(s, x) = x \cdot y - sk^*(y), \quad \alpha = \partial_s \phi, \quad \beta = \nabla\phi, \qquad (17)$$

where $y \in \mathbb{R}^d$ will be chosen later. ¿From definition (14), we deduce, with such a guess,

$$\inf K \geq \phi(1, X_1) - \phi(0, X_0) = (X_1 - X_0) \cdot y - k^*(y),$$

for all $y \in \mathbb{R}^d$. Optimizing in y and using that

$$k(x) = \sup_{y \in \mathbb{R}^d} x \cdot y - k^*(y),$$

we get

$$\inf K \geq k(X_1 - X_0),$$

i.e. the reverse inequality of (12), which concludes the proof.

2.2 Extension to probability measures

The main advantage of the concept of generalized geodesics (ρ, E) as minimizers of $K(\rho, E)$ subject to (4) is that (ρ, E) can achieve boundary data

$$\rho(s = 0, \cdot) = \rho_0, \quad \rho(s = 1, \cdot) = \rho_1, \qquad (18)$$

that are (Borel) probability measures defined on the subset D. Probability measures should be seen in this context, as generalized (or fuzzy) points.

Theorem 2.2. *Let (ρ_0, ρ_1) a pair of probability measures on D. Then $\inf K$ is always finite and does not differ from the Monge-Kantorovich generalized distance between ρ_0 and ρ_1 usually defined by*

$$I_k(\rho_0, \rho_1) =: \inf \int_{D^2} k(x - y)d\mu(x, y), \qquad (19)$$

where the infimum is performed on all nonnegative measures μ on $D \times D$ with projections ρ_0 and ρ_1 on each copy of D.

The relationship with the MKP has been established in [BB] for numerical purposes.

Proof

The proof requires the following fact, known as Kantorovich duality [36], [GM] :

$$I_k(\rho_0, \rho_1) = \sup \int_D (\phi_1(x)d\rho_1(x) - \phi_0(x)d\rho_0(x)), \qquad (20)$$

where ϕ_1 and ϕ_0 are continuous functions on D subject to

$$\phi_1(y) \leq k(x - y) + \phi_0(x), \quad \forall x, y \in D. \qquad (21)$$

¿From definition (19), there is always a minimizer μ, so that

$$\int_{D^2} k(b - a)d\mu(a, b) = I_k(\rho_0, \rho_1).$$

Let us introduce, for this μ,

$$\rho(s, x) = \int_{D^2} \delta(x - X(s, a, b))d\mu(a, b), \qquad (22)$$

$$E(s, x) = \int_{D^2} \partial_s X(s, a, b)\delta(x - X(s, a, b))d\mu(a, b), \qquad (23)$$

where

$$X(s, a, b) = (1 - s)a + sb. \qquad (24)$$

Just as in the proof of Theorem 2.1, compatibility condition (4) is satisfied and, by Jensen's inequality,

$$\int_{D^2} k(b - a)d\mu(a, b) \geq K(\rho, E) \qquad (25)$$

$$\geq \inf K \geq \sup_\phi \int_D (\phi(1, x) d\rho_1(x) - \phi(0, x) d\rho_0(x)),$$

where

$$\partial_s \phi + k^*(\nabla \phi) \leq 0. \tag{26}$$

So we can choose ϕ to be any solution of the Hamilton-Jacobi equation (16) on D. Using the Hopf formula to solve (16) (see [Ba], [Li]), we get

$$\phi(s, x) = \inf_{y \in D} (\phi(0, x + s(y - x)) + sk(y - x)),$$

for all $s \geq 0$. Thus, from (25), we finally get

$$I_k(\rho_0, \rho_1) = \int_{D^2} k(b-a) d\mu(a, b) \geq \inf K \geq \sup_\phi \int_D (\phi(1, x) d\rho_1(x) - \phi(0, x) d\rho_0(x))$$

where

$$\phi(1, x) = \inf_{y \in D} (\phi(0, y) + k(y - x)).$$

Thus, we conclude, using Kantorovich duality (20), that there is no difference between $I_k(\rho_0, \rho_1)$ and $\inf K$, which concludes the proof.

2.3 A decomposition result

¿From the proof of Theorem 2.2, we immediately get the following *decomposition* result that asserts that generalized geodesics are mixtures of classical geodesics.

Theorem 2.3. *Each pair (ρ_0, ρ_1) of (Borel) probability measures on D admits a generalized geodesic (ρ, E) linking them with the following structure*

$$\rho(s, x) = \int_{D^2} \delta(x - X(s, a, b)) d\mu(a, b), \tag{27}$$

$$E(s, x) = \int_{D^2} \partial_s X(s, a, b) \delta(x - X(s, a, b)) d\mu(a, b), \tag{28}$$

where $X(\cdot, a, b)$ is the shortest path between a and b in D

$$X(s, a, b) = (1 - s)a + sb, \tag{29}$$

and μ is a probability measure on D^2 with projections ρ_0 and ρ_1 on each copy of D.

Remark

Under strict convexity assumptions on k, the structure theorem can be made more precise, because of the well known properties of the Monge-Kantorovich problem. Indeed, in such cases, there is a unique minimizer μ with structure

$$\mu(a,b) = \delta(b - T(a))$$

where $T : D \to D$ is a Borel map [GM], [36]. In particular, as $k(x) = |x|^2/2$, [Br1], [Ca], T is a map with Lipschitz convex potential Ψ. (An interesting application of this fact to pure analysis can be found in [Bt].) This potential solves the Monge-Ampère equation

$$\det(D^2\Psi(x))\rho_1(\nabla\Psi(x)) = \rho_0(x) \tag{30}$$

in the weak sense that ρ_1 is the image measure of ρ_0 by $\nabla\Psi$. If D is stricltly convex with a smooth boundary and if ρ_0 and ρ_1 are smooth functions bounded away from zero on D, then Ψ inherits the regularity of the data and becomes a classical solution to the Monge-Ampère equation, as shown by Caffarelli [Ca]. Notice that the assumption that D is a convex set, which is convenient but not at all essential for the existence and uniqueness theory, is crucial for the regularity theory, as pointed out by Caffarelli.

2.4 Relativistic MKT

Beyond the most important cost functions, namely $k(v) = |v|$, which corresponds to the original Monge problem, and $k(v) = |v|^2/2$, which corresponds to the Monge-Ampère equations and is related to PDEs as different as the Euler equations of incompressible flows [Br1], [Br2] and the heat equation [JKO], more general cost functions have been considered in the litterature (see for instance [GM]). Surprisingly, two important cost functions have been neglected, in spite of their obvious geometric and relativistic flavour, namely

$$k(v) = \left(1 - \sqrt{1 - \frac{|v|^2}{c^2}}\right)c^2, \tag{31}$$

(with value $+\infty$ as $|v| > c$) and its dual function

$$\left(\sqrt{1 + \frac{|p|^2}{c^2}} - 1\right)c^2, \tag{32}$$

where $c > 0$ can be interpreted as a maximal speed. Notice that the later interpolates the important cost functions $k(p) = |p|$ and $k(p) = |p|^2/2$ as c varies from 0 to $+\infty$. Certainly, the case of (31) is the more interesting. Indeed, due to the finite maximal propagation speed, the MK problem may have no solution with finite cost, as the support of data ρ_0 and ρ_1 are too far from each other. The most interesting and challenging case is when only a part of the mass can be transported, which seems a very realistic approach to many applications. This case can be easily rephrased as a free boundary optimal transportation problem. ¿From the Analysis point of view, the regularity of the free boundary is certainly a challenging problem.

2.5 A relativistic heat equation

As an application of the relativistic cost (31), let us compute, a relativistic heat equation, defined, in the spirit of [JKO], as a gradient flow of the Boltzmann entropy for the metric corresponding to cost (31). The Boltzmann entropy is given by

$$\eta(\rho) = \int_{\mathbb{R}^d} (\log \rho(x) - 1)\rho(x)dx, \tag{33}$$

where ρ is a density function defined on $\mathbb{R}^d$. To do the computation, we follow the time discrete approach of [JKO], rather than the stricter formalism of [Ot2]. The time dependent solution $\rho(t, x)$ is approximated at each time step $n\delta t$, $n = 1, 2, 3, ...$ by $\rho_n(x)$ subject to achieve

$$\inf_{\rho_n} \left(\nu\eta(\rho_n) + \int_0^1 \int_{\mathbb{R}^d} k(e(s, x))\rho(s, x)dsdx \right),$$

where (ρ, e) are subject to (4) (with $E = \rho e$) and boundary conditions

$$\rho(0, \cdot) = \rho_{n-1}, \quad \rho(1, \cdot) = \rho_n.$$

Here $\nu > 0$ is a parameter. This minimization problem can be easily written as a saddle point problem

$$\inf_{\rho_n} \sup_{\phi} \int_0^1 \int (k(e) - \partial_s\phi - e \cdot \nabla\phi) \rho dsdx$$

$$+ \int (\phi(1, x) + \nu(\log \rho_n(x) - 1))\rho_n(x)dx - \int \phi(0, x)\rho_{n-1}(x)dx,$$

which leads to the optimality conditions

$$\nabla k(e) = \nabla\phi, \quad \phi(1, \cdot) + \nu \log \rho_n = 0,$$

and

$$\partial_s\phi + k^*(\nabla\phi) = 0.$$

The first condition is equivalent to

$$e = \nabla k^*(\nabla \phi) = \frac{\nabla \phi}{\sqrt{1 + \frac{|\nabla \phi|^2}{c^2}}},$$

which, as expected, is always bounded by c. Letting formally δt go to zero, leads to the closure relation

$$\phi = -\nu \log \rho,$$

which, combined to (4), gives the desired relativistic heat equation with propagation speed bounded by c :

$$\partial_t \rho = \nu \nabla \cdot \frac{\rho \nabla \rho}{\sqrt{\rho^2 + \frac{\nu^2 |\nabla \rho|^2}{c^2}}}. \tag{34}$$

Notice that ν has the dimensionality of a kinematic viscosity (length^2/time). This equation interpolates the regular heat equation (as $c \to +\infty$) and the following limit equation, where the propagation speed is always c

$$\partial_t \rho = c \nabla \cdot \rho \frac{\nabla \rho}{|\nabla \rho|}. \tag{35}$$

An interesting output of equation (34) is the concept of 'relativistic Fischer information' defined as the entropy production, namely :

$$\frac{d}{dt} \int \rho \log \rho dx = \int \nu \frac{|\nabla \rho|^2}{\sqrt{\rho^2 + \frac{\nu^2 |\nabla \rho|^2}{c^2}}},$$

which interpolates the classical Fischer information (see [OV] for instance)

$$\nu \int \frac{|\nabla \rho|^2}{\rho} dx$$

and the 'total variation'

$$c \int |\nabla \rho| dx.$$

Let us finally mention that this relativistic heat equation can probably be found among the various "flux limited diffusion" equations used in the theory of radiation hydrodynamics [MM]. Indeed, A very similar equation

$$\partial_t \rho = \nu \nabla \cdot \frac{\rho \nabla \rho}{\rho + \frac{\nu |\nabla \rho|}{c}} \tag{36}$$

can be read (in our notations!) in [MM] (p.479). The author is grateful to Bruno Després for this comment.

2.6 Laplace's equation and Moser's lemma revisited

The formulation of the MKT in terms of generalized geodesics allows us to introduce a genuine extension of the MKT by relaxing the constraint for functional $K(\rho, E)$ to be homogeneous of degree one in the pair (ρ, E), as enforced by dual definition (10), (11). In particular, we can consider functionals that *do not* depend on ρ, such as

$$K(\rho, E) = K(E) = \int_0^1 \int_D k(E(s, x)) ds dx, \qquad (37)$$

at least when E is absolutely continuous with respect to the Lebesgue measure, where k is a fixed continuous convex function on $\mathbb{R}^d$. We can give a more precise definition of K by setting

$$K(\rho, E) = \sup_\beta \int_0^1 \int_D (\beta(s, x) \cdot dE(s, x) - k^*(\beta(s, x)) ds dx), \qquad (38)$$

where β is any continuous function on $[0, 1] \times D$ and k^* is the Legendre-Fenchel transform of k.

Notice that, with this definition, the finiteness of $K(\rho, E)$ implies that E is absolutely continuous with respect to the Lebesgue measure, provided k and k^* are strictly convex (which rules out $k(x) = |x|$ for example). Given probability measures ρ_0, ρ_1 on D, let us find a generalized geodesics between them for this new type of functionals (inaccessible by the usual MKT). The minimization problem amounts to solve

$$\inf_{\rho, E} \sup_\phi \int_0^1 \int_D (k(E) - \partial_s \phi \rho - E \cdot \nabla \phi) + \int_D (\phi(1, \cdot) \rho_1 - \phi(0, \cdot) \rho_0).$$

The saddle-point conditions obtained by differientating in E and ρ are :

$$(\nabla k)(E) = \nabla \phi, \quad \partial_s \phi = 0,$$

which reduces to

$$E(s, x) = E(x) = (\nabla k^*)(\nabla \phi(x)), \quad \rho(s, x) = \rho_0(x)(1 - s) + \rho_1(x) s.$$

In a complete contrast with the MKT, here the data ρ_0 and ρ_1 are just linearly interpolated, while potential ϕ does not depend on s. Because of the compatibility condition

$$\partial_s \rho + \nabla \cdot E = 0,$$

we deduce that ϕ solves the $k-$ Laplace equation

$$-\nabla.((\nabla k^*)(\nabla \phi)) = \rho_1 - \rho_0, \qquad (39)$$

with $k-$ Neumann boundary condition

$$(\nabla k^*)(\nabla \phi)) \cdot n = 0,$$

along ∂D, where n denotes the outward normal to ∂D. To get a more rigorous argument, we notice that to each admissible pair $(\overline{\rho}, \overline{E})$ we may associate a new admissible pair (ρ, E) such that

$$E(s, x) = E(x) \quad \rho(s, x) = \rho_0(x)(1 - s) + \rho_1(x)s,$$

with $K(E) \le K(\overline{E})$, just by setting

$$E(x) = \int_0^1 \overline{E}(s, x)ds.$$

Indeed, K is lowered by Jensen's inequality and (6) is enforced as soon as

$$\int_D \nabla \Phi(x) \cdot E(x)dx = \int \Phi(x)(\rho_1(x) - \rho_0(x))dx \qquad (40)$$

which means, in weak sense that

$$-\nabla \cdot E = \rho_1 - \rho_0$$

with E parallel to ∂D. It follows that

$$\inf K = \inf_E \int_D k(E(x))dx,$$

where E is subject to (1). This new minimization problem is nothing but the dual formulation of the $p-$ Laplacian equation with homogeneous $p-$ Neumann boundary condition. So, we have obtained an interpretation of $p-$ Laplace problems as generalized Monge-Kantorovich problems to the expense of using a cost K which is not homogeneous of degree one. This interpretation may look artificial, since, after all, the interpolation variable s has disappeared at the end as well as the transportation framework. Nevertheless, thanks to this approach, we get a new interpretation, in terms of generalized MKT, of the Moser lemma in its simplest form (see [DM], [GY] for more sophisticated versions). The purpose of Moser's lemma is to construct a smooth map T transporting two given probability densities ρ_0 to ρ_1 that are assumed to be smooth and bounded away from zero on a nice domain D. The simplest Moser construction consists first in solving the Laplace equation

$$-\Delta \phi = \rho_1 - \rho_0, \qquad (41)$$

inside D, with homogeneous Neumann boundary conditions along ∂D, next in introducing a velocity field

$$e(s, x) = \frac{\nabla \phi(x)}{\rho(s, x)},$$

where

$$\rho(s, x) = \rho_0(x)(1 - s) + \rho_1(x)s,$$

which automatically satisfies

$$\partial_s \rho + \nabla \cdot (\rho e) = 0, \quad \rho(0, \cdot) = \rho_0, \quad \rho(1, \cdot) = \rho_1. \tag{42}$$

Then T is obtained, after integrating e, as $T(x) = X(1, x)$ where

$$\partial_s X(s, x) = e(s, X(s, x)), \quad X(0, x) = x.$$

This scheme exactly fits our generalized MK problem with cost

$$K(\rho, E) = K(E) = \frac{|E|^2}{2}.$$

Thus, the Moser construction, in its simpler version, can be interpreted as the solution to a generalized MK problem.

3 Generalized Harmonic functions

3.1 Classical harmonic functions

Let U be a smooth bounded open set in $\mathbb{R}^m$ and let D be the closure of a bounded convex open subset of $\mathbb{R}^d$. A map $u \in \overline{U} \to X(u)$ valued in the interior of D is harmonic if it minimizes

$$\int_U \frac{1}{2} |\nabla X(u)|^2 du \tag{43}$$

among all other maps assuming the same values along the boundary ∂U.
This means that X solves the homogeneous Laplace equation

$$\Delta X = 0. \tag{44}$$

To each map X, we can associate a pair of measures (ρ, E), valued in $\mathbb{R}^+ \times \mathbb{R}^{md}$, defined by

$$\rho(u, x) = \delta(x - X(u)), \quad E(u, x) = \nabla X(u)\delta(x - X(u)), \tag{45}$$

or, more precisely,

$$E_{i\alpha}(u, x) = \partial_{u_\alpha} X_i(u)\delta(x - X(u)), \quad \alpha = 1, ..., m, \quad i = 1, ..., d,$$

for all $(u, x) \in \overline{U} \times D$. These measures satisfy the following compatibility condition in the sense of distributions

$$\nabla_u \rho + \nabla_x \cdot E = 0, \tag{46}$$

with boundary conditions

$$\rho(u, \cdot) = \delta(x - X(u)), \quad \forall u \in \partial U. \tag{47}$$

Equation (46) is a compact notation for

$$\partial_{u_\alpha} \rho + \sum_{i=1,d} \partial_{x_i} E_{i\alpha}, \quad \alpha = 1, ..., m.$$

Equation (46) and boundary condition (47) can be expressed in weak form :

$$\int_{U \times D} (\nabla_u \cdot \phi(u, x) d\rho(u, x) + \nabla_x \phi(u, x) \cdot dE(u, x)) \tag{48}$$

$$= \int_{\partial U \times D} \phi(u, x) \rho(u, dx) \cdot dn(u),$$

for all smooth functions $\phi = (\phi_\alpha, \alpha = 1, ..., m)$ defined on in $\overline{U} \times D$ and valued in $\mathbb{R}^m$, where $dn(u) = n(u) d\mathbf{H}^{m-1}(u)$, $n(u)$ is the outward normal to ∂U at u and $\mathbf{H}^{m-1}$ stands for the $m-1$ dimensional Hausdorff measure. This equation is a compact notation for

$$\int_{U \times D} \left(\sum_{\alpha=1}^{m} \partial_{u_\alpha} \phi_\alpha(u, x) d\rho(u, x) + \sum_{\alpha=1}^{m} \sum_{i=1}^{d} \partial_{x_i} \phi_\alpha(u, x) dE_{i\alpha}(u, x) \right)$$

$$= \int_{\partial U \times D} \sum_{\alpha=1}^{m} \phi_\alpha(u, x) n_\alpha(u) \rho(u, dx) d\mathbf{H}^{m-1}(u).$$

We observe that E is absolutely continuous with respect to ρ and has a Radon-Nikodym density $e(u, x)$. By Jensen's inequality,

$$K(\rho, E) =: \frac{1}{2} \int_{U \times D} |e(s, x)|^2 d\rho(s, x) \le \int_{U} \frac{1}{2} |\nabla_u X(u)|^2 du.$$

Functional K can be equivalently defined by Legendre duality as :

$$K(\rho, E) = \sup_{\alpha, \beta} \int_{U \times D} \alpha(u, x) d\rho(u, x) + \beta(u, x) \cdot dE(u, x), \tag{49}$$

where the supremum is performed over all pair (α, β) of continuous functions defined on $\overline{U} \times D$ respectively valued in $\mathbb{R}$ and $\mathbb{R}^{md}$, subject to satisfy

$$\alpha(u, x) + \sum_{\alpha=1}^{m} \sum_{i=1}^{d} \frac{1}{2} \beta_{i\alpha}(u, x)^2 \le 0 \tag{50}$$

pointwise.

Now, we can define a generalized harmonic functions to be a pair of measures (ρ, E) subject to (46) that minimizes $K(\rho, E)$ defined by (49) as the value of ρ along the boundary ∂U is fixed. This is an obvious generalization of the earlier concept of generalized geodesics which corresponds to the special case $m = 1$, $U =]0, 1[$. We have again a consistency result

Theorem 3.1. *Let* $X : \overline{U} \to \mathbb{R}^d$ *be harmonic, with values in the interior of* D. *Then, the corresponding pair of measures* (ρ, E), *defined on* $\overline{U} \times D$ *by* *(45), achieves the infimum of* K, *defined by (49), among all pairs of measures* (ρ, E) *satisfying (46) with boundary values*

$$\rho(u, x) = \delta(x - X(u)), \quad \forall u \in \partial U. \tag{51}$$

Proof

The proof is very similar to the one we had for generalized harmonic functions. Almost identically, we get

$$\int_U \frac{1}{2} |\nabla_u X(u)|^2 du \geq \inf K$$

$$\geq \sup_\phi \int_{\partial U} \phi(u, X(u)) \cdot dn(u),$$

where $\phi = (\phi_\alpha, \alpha = 1, ..., m)$ is subject to

$$\nabla_u \cdot \phi + \frac{1}{2} |\nabla_x \phi|^2 \leq 0. \tag{52}$$

Let us look for a solution ϕ of the generalized Hamilton-Jacobi equation

$$\nabla_u \cdot \phi + \frac{1}{2} |\nabla_x \phi|^2 = 0, \tag{53}$$

which is linear in x (which turns out to be sufficient for our purpose). We set

$$\phi_\alpha(u, x) = \sum_{i=1}^d w_{i\alpha}(u) x_i + z_\alpha(u), \quad \alpha = 1, ..., m,$$

where w is a smooth fixed function and z is chosen so that

$$\nabla_u.z(u) + \frac{1}{2} |w(u)|^2 = 0,$$

which is always possible (by solving an inhomogeneous Laplace equation

$$-\Delta\zeta(u) = \frac{1}{2}|w(u)|^2, \quad u \in U,$$

and setting $z(u) = \nabla\zeta(u)$. Thus,

$$\int_{\partial U} \phi(u, X(u)) \cdot dn(u) = \int_U \nabla_u \cdot (\phi(u, X(u)) du$$

(by Green's formula)

$$= \int_U \nabla_u \cdot (w(u) \cdot X(u) + z(u)) du = \int_U (\nabla_u \cdot (w(u) \cdot X(u)) - \frac{1}{2}|w(u)|^2) du.$$

So, if we choose

$$w(u) = \nabla_u X(u),$$

we obtain

$$\int_{\partial U} \phi(u, X(u)) \cdot dn(u) = \int_U \left(\nabla_u \cdot (\nabla_u X(u) \cdot X(u)) - \frac{1}{2}|\nabla_u X(u)|^2\right) du,$$

$$= \int_U (\Delta_u X(u) \cdot X(u)) + |\nabla_u X(u)|^2 - \frac{1}{2}|\nabla_u X(u)|^2) du,$$

which is exactly

$$\int_U \frac{1}{2}|\nabla_u X(u)|^2 du,$$

since X is harmonic. Thus, we have obtained the desired reverse inequality

$$\inf K \geq \int_U \frac{1}{2}|\nabla_u X(u)|^2 du,$$

which completes the proof.

3.2 Open problems

Optimality equations

Because of the definition of generalized harmonic functions as minimizers of a (lower semi continuous) convex functional on a compact set of measures, we immediately get the following result.

Proposition 3.1. *Let (A, da) be a probability space. Let $(u, a) \in U \times A \rightarrow X(u, a) \in \mathbb{R}^d$ be a (measurable) family of maps such that*

$$\int_U \int_A |\nabla_u X(u, a)|^2 du\, da < +\infty.$$

Define

$$\bar{\rho}(u,x) = \int_A \delta(x - X(u,a))da.$$

Then there is always a generalized harmonic functions (ρ, E) such that $\rho = \bar{\rho}$ along the boundary of U.

Indeed, it is enough to notice that $\bar{\rho}$, together with

$$\bar{E}(u,x) = \int_A \nabla_u X(u,a)\delta(x - X(u,a))da,$$

defines an admissible solution with finite energy (by Jensen's inequality). Thus, there is an optimal solution by a standard compactness argument. It is harder to establish optimality equations. The formal equations are easily derived as saddle point equations for

$$\inf_{(\rho,E)} \sup_{\phi} \int_{U \times D} \left(\frac{1}{2}|e|^2 - \nabla_u \cdot \phi - \nabla_x \phi \cdot e \right) d\rho$$

(where the boundary terms have been dropped since they do not affect the local equations), namely

$$e = \nabla_x \phi, \quad \nabla_u \cdot \phi + \frac{1}{2}|\nabla_x \phi|^2 = 0. \tag{54}$$

This can equivalently expressed by

$$\partial_j e_{i\alpha} = \partial_i e_{j\alpha} \tag{55}$$

(which means that $e(u,x)$ is curl free in x) and

$$\sum_{\alpha=1}^{m} \partial_\alpha e_{i\alpha} + \sum_{\alpha=1}^{m} \sum_{j=1}^{d} e_{j\alpha} \partial_j e_{i\alpha} = 0 \tag{56}$$

(which is obtained by differentiating in x the second optimality condition). Equation (56) can be written in conservation form, using (46),

$$\sum_{\alpha=1}^{m} \partial_\alpha (\rho e_{i\alpha}) + \sum_{\alpha=1}^{m} \sum_{j=1}^{d} \partial_j (\rho e_{j\alpha} e_{i\alpha}) = 0. \tag{57}$$

Following the techniques of [Br3], it seems possible to establish rigorously the later equation for all generalized harmonic functions. However, it seems difficult to justify the curl-free condition (55).

Superharmonicity of the Boltzmann entropy

An interesting output of the optimality equations is the (formal) superharmonicity in $u \in U$ of the Boltzmann entropy. More precisely, given a smooth generalized harmonic functions (ρ, E) satisfying the optimality conditions, the entropy of ρ

$$\eta(u) =: \int_D (\log \rho(x, u) - 1)\rho(x, u)dx, \tag{58}$$

satisfies $\Delta_u \eta \geq 0$. This property is already known for generalized geodesics, as $m = 1$, $U =]0, 1[$ (corresponding to the MKT with quadratic cost), as the 'displacement convexity' of the entropy, following McCann's [Mc1], [OV]. So, the entropy is superharmonic. Thus, by the maximum principle, the maximum of the entropy must be achieved along the boundary of U. If this result (that we are going to establish only for smooth generalized harmonic functions) is correct in full generality, we can expect the following result :

Conjecture 3.1. Let (ρ, E) be a generalized harmonic function. If $\rho(u, \cdot)$ is absolutely continuous with respect to the Lebesgue measure in D for all u along the boundary ∂U, then this property also holds true for all u inside U.

For the proof, (ρ, E) is assumed to be smooth, $\rho > 0$. For simplicity, implicit summation will be performed on repeated indices and notations

$$\partial_\alpha = \frac{\partial}{\partial u_\alpha}, \quad \alpha = 1, ..., m, \quad \partial_i = \frac{\partial}{\partial x_i}, \quad i = 1, ..., d,$$

will be used. We have

$$\Delta_u \eta(u) = \partial_\alpha \int \log \rho \, \partial_\alpha \rho \, dx$$

$$= \partial_\alpha \int \partial_i \rho \, e_{i\alpha} \, dx$$

(using (46) and integrating by part in x)

$$= \int \partial_\alpha \partial_i \rho \, e_{i\alpha} \, dx + \int \partial_i \rho \, \partial_\alpha e_{i\alpha} \, dx$$

$$= - \int \rho \, e_{j\alpha} \, \partial_i \partial_j e_{i\alpha} \, dx - \int \partial_i \rho \, e_{j\alpha} \partial_j \, e_{i\alpha} \, dx$$

(using (56) and again (46))

$$= \int \rho \, \partial_i e_{j\alpha} \, \partial_j e_{i\alpha} \, dx$$

$$= \int \rho \, \partial_j e_{i\alpha} \, \partial_j e_{i\alpha} \, dx \geq 0$$

(using (55)), which shows that η is superharmonic, as announced.

Decomposition of generalized harmonic functions

A natural question concerns the possibility of decomposing a generalized harmonic functions as a mixture of classical harmonic functions.

Problem 3.1. Let (ρ, E) be a generalized harmonic function. Is there a probability space (A, da) and a family $(u, a) \in U \times A \to X(u, a) \in \mathbb{R}^d$ of harmonic functions, i.e.

$$\Delta_u X(u, a) = 0,$$

such that

$$\rho(u, x) = \int_A \delta(x - X(u, a)) da, \quad E(u, x) = \int_A \nabla_u X(u, a) \delta(x - X(u, a)) da$$

holds true?

We already know that the answer is positive as $U =]0, 1[$, $m = 1$, which corresponds to the case of generalized geodesics. It is fairly clear that the answer is also positive as the target is one-dimensional, i.e. as $d = 1$, because of the maximum principle and the irrelevance of the curl free condition (55). Let us sketch a tentative proof, which is far from being complete.

A tentative proof

The idea of the proof, in case $d = 1$, is based on the fact that, given a generalized harmonic function (ρ, E), we can always write

$$\rho(u, x) = \int_A \delta(x - X(u, a)) da, \tag{59}$$

where $A = [0, 1]$ equipped with the Lebesgue measure da and X is nondecreasing in a for each values of $u \in \partial U$. Now, we claim that, because of the monotonicity of X and because (46) is just an ODE in x,
1) we can write

$$E(s, x) = \int_A \nabla_u X(u, a) \delta(x - X(u, a)) da,$$

2) the convexity inequality

$$K(\rho, E) \leq \frac{1}{2} \int_{U \times A} |\nabla_u X(u, a)|^2 da$$

actually is an equality. (This has to be proven with some care.) Then we conclude that, for each a, $X(\cdot, a)$ must be harmonic. Otherwise, we could introduce, for each fixed a, $\overline{X}(\cdot, a)$ to be the harmonic extension in U of the values of $X(\cdot, a)$ along ∂U. Because of the maximum principle, $\overline{X}(u, a)$ is nondecreasing in a. Defining the corresponding $(\overline{\rho}, \overline{E})$, and using that $\overline{X}$ is harmonic, we would get

$$K(\overline{\rho}, \overline{E}) < K(\rho, E),$$

which is a contradiction since (ρ, E) is supposed to be a generalized harmonic function.

4 Multiphasic MKT

As seen earlier, the MKT on a subset D of $\mathbb{R}^d$ (still assumed to be the closure of a bounded convex open set), is just a theory of generalized geodesics, or, equivalently, following Otto's point of view [Ot2], a theory of geodesics on the "manifold" $Prob(D)$ of all probability measures on D. It is therefore natural to extend this idea to more complex (convex) "manifolds". The most interesting case, in our opinion, is the the set $DS(D)$ of all doubly stochastic probability measures on D, namely the set of all Borel measures μ on $D \times D$ having as projection on each copy of D the (normalized) Lebesgue measure on D, which means

$$\int_{D \times D} f(x) d\mu(x, y) = \int_{D \times D} f(y) d\mu(x, y) = \int_{D} f(x) dx,$$

for all continuous functions f on D. As $d > 1$, this compact convex set turns out to be just the weak closure of the group of orientation and volume preserving diffeomorphisms of D, usually denoted by $SDiff(D)$ [AK], [Ne], through the following embedding

$$g \in SDiff(D) \to \mu_g \in DS(D), \quad \mu_g(x, y) = \delta(y - g(x)).$$

This group is of particular importance because it is the configuration space of incompressible fluids. $SDiff(D)$ is naturally embedded in the space $L^2(D, \mathbb{R}^d)$ of all square Lebesgue integrable maps from D to $\mathbb{R}^d$. Therefore, $SDiff(D)$ inherits the L^2 metric. Then, as pointed out by Arnold [AK], the equations of geodesic curves along $SDiff(D)$ exactly are the Euler equations of incompressible inviscid fluids (see also [MP] and [Br2]).

In our framework, it is very easy to define generalized geodesic curves (and even harmonic maps!) on $DS(D)$.

Definition 4.1. *Given μ_0, μ_1 in $DS(D)$, we define a (minimizing) generalized geodesic curve joining μ_0 and μ_1 to be a pair (μ, E) of (Borel) measures defined on $Q = [0, 1] \times D \times D$ and valued in $\mathbb{R}_+ \times \mathbb{R}^d$ such that*

$$\int_Q \partial_s f(s, x, y) d\mu(s, x, y) + \nabla_x f(s, x, y) \cdot dE(s, x, y) \tag{60}$$

$$= \int_{D^2} f(1, x, y) d\mu_1(x, y) - \int_{D^2} f(0, x, y) d\mu_0(x, y),$$

for all smooth function f on $[0, 1] \times D^2$, *and*

$$\int_Q f(s, x) d\mu(s, x, y) = \int_0^1 \int_D f(s, x) dx ds, \tag{61}$$

for all continuous function f on $[0, 1] \times D$, *that minimizes*

$$K(\mu, E) = \sup_{\alpha, \beta} \int_Q \alpha(s, x, y) d\mu(s, x, y) + \beta(s, x, y) \cdot dE(s, x, y) \tag{62}$$

where the supremum is performed over all pair (α, β) *of continuous functions defined on Q respectively valued in* $\mathbb{R}$ *and* $\mathbb{R}^{md}$, *subject to satisfy*

$$\alpha(s, x, y) + \frac{1}{2} |\beta(s, x, y)|^2 \le 0. \tag{63}$$

pointwise.

This can be seen as a multiphasic MK problem, where to each point $y \in D$ is attached a "phase" described by $\mu(\cdot, \cdot, y)$ and $E(\cdot, \cdot, y)$. These phases are coupled by constraint (61) which forces the different phases to share the volume available in D during their motion. Not surprisingly, this makes the optimality equations more subtle than in the classical MKT. Indeed, there is a Lagrange multiplier corresponding to constraint (61) that physically speaking is the pressure $p(s, x)$ of the fluid at each point $x \in D$ and each $s \in [0, 1]$. The formal optimality conditions read

$$E(s, x, y)/\mu(s, x, y) = e(s, x, y), \quad e(s, x, y) = \nabla_x \phi(s, x, y), \tag{64}$$

$$\partial_s \phi(s, x, y) + \frac{1}{2} |\nabla_x \phi(s, x, y)|^2 + p(t, x) = 0.$$

This multiphasic MK problem has been studied in details in [Br3] and related to the classical Euler equations. (In some cases it is shown that generalized geodesics can be approximated by classical solutions to the Euler equations with vanishing forcing.) To motivate further researches, let us just quote two results that are true for any pair of data (μ_0, μ_1) in $DS(D)$. First, ∇p is uniquely defined (although there may be several generalized minimizing geodesic between μ_0 and μ_1). This fact follows easily from convex duality, but is rather surprising from the classical fluid mechanics point of view. Next, ∇p has a (very) limited regularity. It is a locally bounded measures in the interior of $[0, 1] \times D$, which is not obvious and follows from the minimization principle. Further regularity can therefore be expected (maybe the second derivatives in space of p are also measures?), in particular as the data μ_0 and μ_1 are absolutely continuous with respect to the Lebesgue measure on D^2 with smooth positive density, as in Caffarelli's regularity theory of the (quadratic) MKT [Ca]. It is amusing to notice that (at least formally) the Boltzmann entropy, here defined by

$$\eta(\mu) = \int_{D^2} (\log \mu(x,y) - 1)\mu(x,y)dxdy, \qquad (65)$$

is again "displacement convex" along generalized geodesics. The formal calculation is almost identical to those previously performed in these notes. But this has not been rigorously proven so far.

5 Generalized extremal surfaces

In this section, we first consider a (hyper)surface Σ of dimension m embedded in $\mathbb{R}^d$. We assume that Σ is the image $\Sigma = X(U)$ of a nice domain U in $\mathbb{R}^m$ by a smooth map X with values in a convex subdomain $\mathbf{D}$ of $\mathbb{R}^d$.
For each sequence $i = (i_1, ..., i_m)$ of integers such that $1 \le i_1 < ... < i_m \le \mathbf{d}$, we associate to X a measure $\rho_i(x)$ defined by

$$\rho_i(x) = \int_U \delta(x - X(u)) \sum_\sigma \epsilon(\sigma)\partial_{\sigma_1} X_{i_1}(u)...\partial_{\sigma_m} X_{i_m}(u)du, \qquad (66)$$

where σ is any permutation of the m first integers and $\epsilon(\sigma)$ denotes its signature. For each σ and each $i = (i_1, ..., i_m)$ in $\{1, ..., \mathbf{d}\}^m$, we set

$$\rho_{\sigma_i} = \epsilon(\sigma)\rho_i,$$

so that from now on ρ_i is antisymmetric in i. If

$$f = \sum_{1 \le i_1 < ... < i_m \le \mathbf{d}} f_i(x)dx_{i_1} \wedge .... \wedge dx_{i_\mathbf{d}}$$

is a differential form of degree m on $\mathbb{R}^d$, $\rho = (\rho_i, \ i_1 < ... < i_m)$ acts as a current (i.e. a linear form on differential forms, see [GMS] for instance) on f by the duality bracket

$$< \rho, f >=: \sum_i < \rho_i, f_i > \qquad (67)$$

$$= \sum_i \sum_\sigma \epsilon(\sigma) \int_U (f_{i_1,...i_m})(X(u))\partial_{\sigma_1} X_{i_1}(u)...\partial_{\sigma_m} X_{i_m}(u)du,$$

which (by the area formula) is nothing but the integral of f on Σ. If f is the derivative of a $m-1$ differential form ϕ, we get from the Stokes theorem that

$$\int_\Sigma d\phi = \int_{\partial\Sigma} \phi,$$

i.e.

$$\sum_i \sum_\sigma \epsilon(\sigma) \int_U (\partial_{i_1}\phi_{i_2,\dots i_m})(X(u))\partial_{\sigma_1} X_{i_1}(u)\dots\partial_{\sigma_m} X_{i_m}(u)du = \int_{\partial\Sigma} \phi, \quad (68)$$

which implies, in the distributional sense,

$$\sum_{i_1=1}^{\mathbf{d}} \partial_{i_1}\rho_{i_1,\dots i_m} = 0, \quad (69)$$

for all $1 \le i_2 < \dots < i_m \le \mathbf{d}$. The Euclidean area of the embedded surface Σ is given [EGr] by

$$\int_U \sqrt{\sum_i \left(\sum_\sigma \epsilon(\sigma)\partial_{\sigma_1} X_{i_1}(u)\dots\partial_{\sigma_m} X_{i_m}(u)\right)^2} \, du, \quad (70)$$

which can be written in terms of ρ as

$$\int_{\mathbf{D}} k(\rho), \quad k(\rho) = \sqrt{\sum_i \rho_i^2} \quad (71)$$

or, by duality,

$$\sup_f < \rho, f > \quad (72)$$

where the supremum is performed over all compactly supported differential form $f = (f_i)$ of degree m in $\mathbb{R}^{\mathbf{d}}$ such that

$$\sum_i f_i(x)^2 \le 1, \quad \forall x \in \mathbb{R}^{\mathbf{d}}.$$

We can now define a generalized surface and a generalized minimal surface :

Definition 5.1. Let $\rho = (\rho_i)$ be a family, antisymmetric in $i \in \{1,\dots,\mathbf{d}\}^m$, of measures ρ_i defined on a subset of $\mathbb{R}^{\mathbf{d}}$, denoted by $\mathbf{D}$ (and assumed to be the closure of a bounded open convex set). We say that ρ is a generalized surface lying in $\mathbf{D}$ if

$$\sum_i \int_{\mathbf{D}} \partial_{i_1}\phi_{i_2,\dots i_m}(x)d\rho_i(x) = 0, \quad (73)$$

for each family of smooth function ϕ_i, antisymmetric in i and compactly supported in the interior of $\mathbf{D}$.

Definition 5.2. *Let k be a nonnegative continuous convex function, homogeneous of degree one. We say that a generalized surface ρ in $\mathbf{D}$ has a finite $k-$ area if*

$$K(\rho) = \int_{\mathbf{D}} k(\rho) < +\infty.$$

We say that ρ is a $k-$ generalized minimal surface if

$$K(\rho) \leq K(\bar{\rho})$$

for every generalized surface $\bar{\rho}$ lying in $\mathbf{D}$ such that

$$\sum_i \int_{\mathbf{D}} \partial_{i_1} \phi_{i_2,\ldots i_m}(x)(d\rho_i(x) - d\bar{\rho}_i(x)), \tag{74}$$

for each family of smooth function ϕ_i defined on $\mathbb{R}^d$, antisymmetric in i.

Notice that (74) plays the role of a boundary condition along $\partial \mathbf{D}$ (this is why the ϕ_i are not supposed to vanish along the boundary!).

5.1 MKT revisited as a subset of generalized surface theory

The MKT corresponds to the particular case when

$$m = 1, \quad \mathbf{D} = D \times [0,1], \quad \mathbf{d} = d+1.$$

A current point of $\mathbf{D}$ is denoted by

$$\mathbf{x} = (x_1, \ldots, x_d, s) = (x, s),$$

and, accordingly, $\rho_i(\mathbf{x})$ and $\rho_s(\mathbf{x})$ respectively correspond to $E_i(s, x)$ and $\rho(s, x)$, for $i = 1, \ldots, d$ in our previous notations. Similarly $k(\rho)$ is now $k(\rho, E)$. The boundary values $\bar{\rho}$ are null along the $[0,1] \times \partial D$ and, with the previous notations, they are given by ρ_0 on $\{0\} \times D$ and ρ_1 on $\{1\} \times D$.

5.2 Degenerate quadratic cost functions

Since the MKT has been extended to cost functions $k(\rho, E)$ that are not homogeneous of degree one, we may extend the generalized surface theory in the same way. Not surprisingly, in the special case

$$\mathbf{D} = D \times [0,1], \quad \mathbf{d} = d+1,$$

where current points of $\mathbf{D}$ are denoted by

$$\mathbf{x} = (x_1, \ldots, x_d, s) = (x, s),$$

the (degenerate) quadratic cost

$$k(\rho) = \sum_{1 \le i_1 < ... < i_m < s} \rho_i^2$$

(where the ρ_i for which $i_m = s$ are absent) will directly lead to a Hodge-Laplace problem for differential forms of degree $m - 1$ in D.
For example, if

$$m = 2, \quad \mathbf{D} = D \times [0, 1], \quad \mathbf{d} = 5 + 1,$$

and a point is denoted by (t, x_1, x_2, x_3, s) we recover the (elliptic) Maxwell equations in $\mathbb{R}^4$: In classical notations

$$-\nabla \cdot E = \rho_1 - \rho_0, \quad \partial_t E + \nabla \times B = J_1 - J_0, \tag{75}$$

$$\nabla \cdot B = 0, \quad \partial_t B + \nabla \times E = 0. \tag{76}$$

Because of the degeneracy of the cost function, the "electromagnetic field" (E, B) depends only the classical time space variable (t, x), while the "charge and current" densities (ρ, J) are linear interpolation in s of the "boundary" data (ρ_0, J_0) given at $s = 0$ and (ρ_0, J_0) at $s = 1$:

$$\rho(t, s, x) = (1 - s)\rho_0(t, x) + s\rho_1(t, x), \quad J(t, s, x) = (1 - s)J_0(t, x) + sJ_1(t, x). \tag{77}$$

(This will be checked in more details in a subsequent section.)

6 Generalized extremal surfaces in $\mathbb{R}^5$ and Electrodynamics

In order to address the framework of Electrodynamics, we substitute for the Euclidean metric of $\mathbb{R}^\mathbf{d}$ the Minkowski metric with signature $(-, +, ..., +)$. Assume that $m = 2$ and $\mathbf{d} = 5$. A point in $\mathbb{R}^\mathbf{d}$ will be denoted (t, x_1, x_2, x_3, s) and the partial derivative will be denoted accordingly (i.e. $\partial_t, \partial_1, ..., \partial_s$). The components of ρ are denoted

$$\rho_{st} = \rho,$$

$$\rho_{it} = E_i,$$

$$\rho_{is} = -J_i,$$

$$\rho_{12} = B_3, \quad \rho_{23} = B_1, \quad \rho_{31} = B_2.$$

Compatibility conditions (69) become

$$\partial_t \rho + \nabla \cdot J = 0 \tag{78}$$

(here $\cdot$ is the inner product in $\mathbb{R}^3$),

$$\partial_s \rho + \nabla \cdot E = 0, \tag{79}$$

$$\partial_t E - \partial_s J + \nabla \times B = 0. \tag{80}$$

We consider a functional K defined by

$$K(\rho, J, E, B) = \int k(\rho, J, E, B), \tag{81}$$

where k is a given function defined on $\mathbb{R}^{10}$. For example, the Euclidean area corresponds to

$$k(\rho, J, E, B) = \sqrt{\rho^2 + J^2 + E^2 + B^2}, \tag{82}$$

while the Minkowski area is given by

$$k(\rho, J, E, B) = \sqrt{\rho^2 - J^2 + E^2 - B^2}. \tag{83}$$

6.1 Recovery of the Maxwell equations

In the same way as the Laplace equation can be recovered from the MKT by using a degenerate quadratic functional, the Maxwell equations can be easily recovered by using a simplified functional such as

$$k(\rho, J, E, B) = \frac{E^2 - B^2}{2}. \tag{84}$$

To check this statement, we introduce two Lagrange multipliers $\phi(t, s, x) \in \mathbf{R}$, $A(t, s, x) \in \mathbf{R}^3$ to enforce the compatibility conditions and we define the corresponding Lagrangian $L(\rho, J, E, B, \phi, A)$:

$$\int \{\frac{E^2 - B^2}{2} - \rho \partial_s \phi - E.\nabla \phi - E.\partial_t A + J.\partial_s A - B.\nabla \times A\} dt ds dx \tag{85}$$

where the boundary terms have been skipped since they do not affect the local equations. Varying the Lagrangian yields

$$\partial_s \phi = 0, \quad \partial_s A = 0, \tag{86}$$

$$E = \partial_t A + \nabla \phi, \quad B = -\nabla \times A. \tag{87}$$

We see that ϕ, A, E, B depend only on (t, x) and not on s. By using the compatibility conditions (79), (80) and eliminating ϕ and A, we get

$$\partial_s \rho + \nabla \cdot E = 0, \tag{88}$$

$$\nabla \cdot B = 0, \tag{89}$$

$$\partial_t E - \partial_s J - \nabla \times B = 0, \tag{90}$$

$$\partial_t B + \nabla \times E = 0. \tag{91}$$

We deduce that ρ and J depend linearly on s, namely

$$\rho(t, s, x) = (1 - s)\rho_0(t, x) + s\rho_1(t, x), \tag{92}$$

$$J(t, s, x) = (1 - s)J_0(t, x) + sJ_1(t, x), \tag{93}$$

and (E, B) satisfy the Maxwell equations

$$-\nabla \cdot E = \rho_1 - \rho_0, \quad \partial_t E - \nabla \times B = J_1 - J_0, \tag{94}$$

$$\nabla \cdot B = 0, \quad \partial_t B + \nabla \times E = 0. \tag{95}$$

6.2 Derivation of a set of nonlinear Maxwell equations

Let us now get the variational equations corresponding to the original, non quadratic, Action (83), with compatibility conditions (79), (80). The corresponding Lagrangian is given by

$$\int \{R - \rho \partial_s \phi - E.\nabla \phi - E.\partial_t A + J.\partial_s A - B.\nabla \times A\} dt\, ds\, dx, \tag{96}$$

where

$$R = \sqrt{\rho^2 + E^2 - J^2 - B^2} \tag{97}$$

and boundary terms have been disregarded since they do not affect the local equations we are looking for, although they play an important role to get correct boundary conditions. Varying the Lagrangian leads to

$$\rho = R\partial_s \phi, \quad J = R\partial_s A, \tag{98}$$

$$E = R(\partial_t A + \nabla \phi), \quad B = -R \nabla \times A. \tag{99}$$

By using compatibility conditions (79), (80) and eliminating the Lagrange multipliers in (98), (99), ϕ and A, we deduce, after elementary calculations,

$$\partial_t (JR^{-1}) - \partial_s (ER^{-1}) + \nabla(\rho R^{-1}) = 0, \tag{100}$$

$$\partial_t (BR^{-1}) + \nabla \times (ER^{-1}) = 0, \quad \nabla \cdot (BR^{-1}) = 0. \tag{101}$$

To get an evolution equation for ρ, we can use compatibility condition (78) and disregard (79). Notice that an additional compatibility condition can also be derived from (98), (99), namely

$$\nabla \times (JR^{-1}) + \partial_s (BR^{-1}) = 0. \tag{102}$$

Of course, all these compatibility conditions are not independent from each other and we can select those which lead to a self-consistent system of time-evolution equations.

So, we retain for the set of variables

$$\rho, \; j = JR^{-1}, \; E, b = BR^{-1}. \tag{103}$$

the following evolution equations

$$\partial_t \rho + \nabla \cdot (Rj) = 0. \tag{104}$$

$$\partial_t j - \partial_s (ER^{-1}) + \nabla(\rho R^{-1}) = 0, \tag{105}$$

$$\partial_t E - \partial_s (Rj) - \nabla \times (Rb) = 0, \tag{106}$$

$$\partial_t b + \nabla \times (ER^{-1}) = 0, \tag{107}$$

where R is now expressed by

$$R = \sqrt{\frac{\rho^2 + E^2}{1 + j^2 + b^2}}. \tag{108}$$

After introducing

$$Z = \sqrt{(\rho^2 + E^2)(1 + j^2 + b^2)} \tag{109}$$

and noticing that

$$\frac{\partial Z}{\partial \rho} = \rho R^{-1}, \quad \frac{\partial Z}{\partial j} = Rj, \quad \frac{\partial Z}{\partial E} = ER^{-1}, \quad \frac{\partial Z}{\partial b} = Rb, \tag{110}$$

we finally get such a system, namely

$$\partial_t \rho = -\nabla.(\frac{\partial Z}{\partial j}), \qquad \partial_t E = \partial_s(\frac{\partial Z}{\partial j}) + \nabla \times (\frac{\partial Z}{\partial b}), \qquad (111)$$

$$\partial_t j = \partial_s(\frac{\partial Z}{\partial E}) - \nabla(\frac{\partial Z}{\partial \rho}), \qquad \partial_t b = -\nabla \times (\frac{\partial Z}{\partial E}). \qquad (112)$$

¿From now on, we call these equations $MKMEs$ (Monge - Kantorovich Maxwell equations).

6.3 An Euler-Maxwell-type system

Since the $MKMEs$ are time evolution equations, it is natural to supplement them with initial value conditions. We also need boundary conditions, at least for the interpolation variable $s \in [0, 1]$. As a matter of fact, the most interesting boundary conditions are

$$(E, B)(t, s = 0, x) = (E, B)(t, s = 1, x) = 0. \qquad (113)$$

Observe that these conditions are natural since, in the special case when (ρ, J, E, B) is generated from a surface X they correspond to

$$\partial_s X(t, s = 0) = \partial_s X(t, s = 1) = 0, \qquad (114)$$

which is the right free boundary condition for an extremal surface.

It is now interesting to focus on the fields (ρ, J) at the end points $s = 0$ and $s = 1$, since they are not prescribed any longer. Let us introduce notations

$$(\rho_-, J_-)(t, x) = (\rho, J)(t, s = 0, x), \quad (\rho_+, J_+)(t, x) = (\rho, J)(t, s = 1, x). \qquad (115)$$

Remarkable simplifications occur in the $MKMEs$ restricted at $s = 0$ and $s = 1$ for such boundary conditions. Indeed, we get, at $s = 0$ and $s = 1$, since $E = B = 0$, from (97), (108),

$$R = \sqrt{\rho^2 - J^2} = \frac{\rho}{\sqrt{1 + j^2}}, \qquad (116)$$

$$\frac{\rho}{R} = \sqrt{1 + j^2}, \quad Rj = \frac{\rho j}{\sqrt{1 + j^2}}, \quad \partial_s(Rj) = \partial_s(\frac{\rho j}{\sqrt{1 + j^2}}). \qquad (117)$$

Thus, using (104), (105), (106), we get at $s = 0$ and $s = 1$

$$\partial_t \rho_- + \nabla.(\rho_- v_-) = \partial_t \rho_+ + \nabla.(\rho_+ v_+) = 0, \qquad (118)$$

where we introduce notation

$$v = \frac{j}{\sqrt{1+j^2}}, \tag{119}$$

$$\partial_t j_- + \nabla\sqrt{1+j_-^2} = \partial_s e_{|s=0}, \quad \partial_t j_+ + \nabla\sqrt{1+j_+^2} = \partial_s e_{|s=1}, \tag{120}$$

$$\partial_s(\frac{\rho j}{\sqrt{1+j^2}})_{|s=0} = \partial_s(\frac{\rho j}{\sqrt{1+j^2}})_{|s=1} = 0. \tag{121}$$

Using the standard identity

$$\nabla\sqrt{1+j^2} = (\frac{j}{\sqrt{1+j^2}}.\nabla)j + \frac{j}{\sqrt{1+j^2}} \times (\nabla \times j), \tag{122}$$

and (102) we can rewrite (120),

$$(\partial_t + v_-.\nabla)j_- = E_- + v_- \times B_-, \tag{123}$$

$$(\partial_t + v_+.\nabla)j_+ = E_+ + v_+ \times B_+, \tag{124}$$

where the 'electromagnetic' fields $(E_-, B_-)(t,x)$, $(E_+, B_+)(t,x)$ are defined by

$$E_-(t,x) = \partial_s e(t, s = 0, x), \quad E_+(t,x) = \partial_s e(t, s = 1, x), \tag{125}$$

$$B_-(t,x) = \partial_s b(t, s = 0, x), \quad B_+(t,x) = \partial_s b(t, s = 1, x). \tag{126}$$

These fields are linked by (107)

$$\partial_t B_- + \nabla \times E_- = \partial_t B_+ + \nabla \times E_+ = 0, \quad \nabla.B_- = \nabla.B_+ = 0, \tag{127}$$

which is one half of the usual Maxwell equations. Thus, with equations (118), (123), (124) and (127), we are not far from the standard (relativistic pressureless) Euler-Maxwell system, for which the very same equations would be supplemented by

$$E_- = -\frac{E_0}{m_-}, \quad E_+ = \frac{E_0}{m_+}, \quad B_- = -\frac{B_0}{m_-}, \quad B_+ = \frac{B_0}{m_+}, \tag{128}$$

$$\partial_t E_0 - \nabla \times B_0 = \rho_+ j_+ - \rho_- j_-, \quad \nabla.E_0 = \rho_+ - \rho_-, \tag{129}$$

where m_- (resp. m_+) denote the mass of negatively (resp. positively) charged particles and $(E_0, B_0)(t,x)$ the electromagnetic field. Of course, in the resulting Euler-Maxwell system, the s variable has completely disappeared. For the $MKMEs$, however, such a simple closure is not possible since we cannot eliminate the interpolation variable s and the coupling is much subtler.

References

[AK] V.I.Arnold, B.Khesin, *Topological methods in Hydrodynamics*, Springer Verlag, 1998.

[Ba] G. Barles, *Solutions de viscosité des équations d'Hamilton-Jacobi*, Mathématiques et applications 17, Springer, 1994.

[Bt] F. Barthe, *On a reverse form of the Brascamp-Lieb inequality*, Invent. Math. *134 (1998) 335-361.*

[BB] J.-D. Benamou, Y. Brenier, A Computational Fluid Mechanics solution to the Monge-Kantorovich mass transfer problem, *to appear in Numerische Math.*

[Br0] Y.Brenier, *Décomposition polaire et réarrangement monotone des champs de vecteurs*, C. R. Acad. Sci. Paris Sér. I Math. *305 (1987), no. 19, 805-808.*

[Br1] Y.Brenier, *Polar factorization and monotone rearrangement of vector-valued functions*, Comm. Pure Appl. Math. 64 (1991) 375-417.

[Br2] Y.Brenier, *Derivation of the Euler equations from a caricature of Coulomb interaction* Comm. Math. Physics 212 (2000) 93-104.

[Br3] Y.Brenier, *Minimal geodesics on groups of volume-preserving maps*, Comm. Pure Appl. Math. 52 (1999) 411-452.

[Ca] L.A. Caffarelli, *Boundary regularity of maps with convex potentials.* Ann. of Math. (2) 144 (1996), no. 3, 453-496.

[CP] M.J. Cullen and R.J. Purser, *An extended lagrangian theory of semigeostrophic frontogenesis*, J. Atmos. Sci., 41:1477-1497, 1984.

[DM] B. Dacorogna, J. Moser, *On a partial differential equation involving the Jacobian determinant*, Ann. Inst. H. Poincaré Anal. Non Linéaire 7 (1990) 1-26.

[EGn] L.C. Evans, W. Gangbo, *Differential equations methods for the Monge-Kantorovich mass transfer problem*, Mem. Amer. Math. Soc. 137 (1999), no. 653.

[EGr] L.C. Evans, Gariepy, R. *Measure theory and fine properties of functions*, Studies in Advanced Mathematics, CRC Press, Boca Raton, FL, 1992.

[GM] W. Gangbo, R. McCann, *The geometry of optimal transportation*, Acta Math. 177 (1996) 113-161.

[GMS] M. Giaquinta, G. Modica, J. Souček, *Cartesian currents in the calculus of variations. II. Variational integrals*, Springer-Verlag, Berlin, 1998.

[JKO] D. Kinderlehrer, R. Jordan, F. Otto, *The variational formulation of the Fokker-Planck equation*, SIAM J. Math. Anal. 29 (1998), no. 1,1-17.

[Ka] L.V. Kantorovich, *On a problem of Monge*, Uspekhi Mat. Nauk. 3 (1948), 225-226.

[KS] M. Knott, C.S. Smith, *On the optimal mapping of distributions*, J. Optim. Theory Appl. 43 (1984), no. 1, 39-49.

[Li] P.-L. Lions, *Generalized solutions of Hamilton Jacobi equations*, Research Notes in Mathematics, 69. Pitman, Boston, Mass.-London, 1982.

[Mc1] R. McCann, *A convexity principle for interacting gases*, Adv. Math. 128 (1997) 153-179.

[Mc2] R. McCann, *Polar factorization of maps on Riemannian manifolds*, Geom. Funct. Anal. 11 (2001) 589-608.

[MP] C. Marchioro, M. Pulvirenti, *Mathematical theory of incompressible nonviscous fluids*, Springer, New York, 1994.

[MM] D. Mihalas, B. Mihalas, *Foundations of radiation hydrodynamics*, Oxford University Press, New York, 1984.

[Ne] Y. Neretin, *Categories of bistochastic measures and representations of some infinite-dimensional groups*, Sb. 183 (1992), no. 2, 52-76.

[Ot1] F. Otto, *Evolution of microstructure in unstable porous media flow: a relaxational approach,*. Comm. Pure Appl. Math. 52 (1999) 873-915.

[Ot2] F. Otto, *The geometry of dissipative evolution equations: the porous medium equation*, Comm. Partial Differential Equations 26 (2001) 101-174.

[OV] F. Otto, C. Villani, *Generalization of an inequality by Talagrand and links with the logarithmic Sobolev inequality*, J. Funct. Anal. 173 (2000) 361-400.

[RR] S.T. Rachev, L. Rüschendorf, *Mass transportation problems*, Vol. I and II. Probability and its Applications, Springer-Verlag, New York, 1998.T

[GY] T. Rivière, D. Ye, *Une résolution de l'équation à forme volume prescrite*, C. R. Acad. Sci. Paris série I Math. 319 (1994) 25-28.

[Su] V.N. Sudakov, *Geometric problems in the theory of infinite-dimensional probability distributions*, Proceedings of the Steklov Institute 141 (1979) 1–178.

[Ta] L. Tartar, *The compensated compactness method applied to systems of conservation laws*. Systems of nonlinear PDE, NATO ASI series, Reidel,Dordecht, 1983.

[Yo] L. C. Young, *Lectures on the calculus of variations*. Chelsea,New York, 1980.

Existence and stability results in the L^1 theory of optimal transportation

Luigi Ambrosio[1] and Aldo Pratelli[2]

[1] Scuola Normale Superiore
  Piazza dei Cavalieri, 56100 Pisa, Italy
  luigi@ambrosio.sns.it
[2] Scuola Normale Superiore
  Piazza dei Cavalieri, 56100 Pisa Italy
  a.pratelli@sns.it

1 Introduction

In 1781, G.Monge raised the problem of transporting a given distribution of matter (a pile of sand for instance) into another (an excavation for instance) in such a way that the work done is minimal. Denoting by h_0, $h_1 : \mathbf{R}^2 \to [0, +\infty)$ the Borel functions describing the initial and final distribution of matter, there is obviously a compatibility condition, that the total mass is the same:

$$\int_{\mathbf{R}^2} h_0(x)\, dx = \int_{\mathbf{R}^2} h_1(y)\, dy. \qquad (1)$$

Assuming with no loss of generality that the total mass is 1, we say that a Borel map $t : \mathbf{R}^2 \to \mathbf{R}^2$ is a *transport* if a local version of the balance of mass condition holds, namely

$$\int_{t^{-1}(E)} h_0(x)\, dx = \int_E h_1(y)\, dy \qquad \text{for any } E \subset \mathbb{R}^2 \text{ Borel}. \qquad (2)$$

Then, the Monge problem consists in minimizing the work of transportation in the class of transports, i.e.

$$\min\left\{ \int_{\mathbf{R}^2} |t(x) - x| h_0(x)\, dx : \ t \text{ transport} \right\}. \qquad (3)$$

The Monge transport problem can be easily generalized in many directions, and all these generalizations have proved to be quite useful:
• General measurable spaces X, Y, with measurable maps $t : X \to Y$;

• General probability measures μ in X and ν in Y. In this case the local balance of mass condition (2) reads as follows:

$$\nu(E) = \mu(t^{-1}(E)) \qquad \text{for any } E \subset Y \text{ measurable.} \qquad (4)$$

This means that the push-forward operator $t_\#$ induced by t, mapping probability measures in X into probability measures in Y, maps μ into ν.

• General cost functions: a measurable map $c : X \times Y \to [0, +\infty]$. In this case the cost to be minimized is

$$W(t) := \int_X c(x, t(x)) \, d\mu(x).$$

The transport problem has by now an impressive number of applications, covering Non-linear PDE's, Calculus of Variations, Probability, Economics, Statistical Mechanics and many other fields. We refer to the surveys/books [3], [23], [36], [43], [44] for more informations on this wide topic.

Even in Euclidean spaces, the problem of existence of optimal transport maps is far from being trivial, mainly due to the non-linearity with respect to t of the condition $t_\#\mu = \nu$. In particular the class of transports is not closed with respect to any reasonable weak topology. Furthermore, it is easy to build examples where the Monge problem is ill-posed simply because there is no transport map: this happens for instance when μ is a Dirac mass and ν is not a Dirac mass.

In order to overcome these difficulties, in 1942 L.V.Kantorovich proposed in [31], [32] a notion of weak solution of the transport problem. He suggested to look for *plannings* instead of transports, i.e. probability measures γ in $X \times Y$ whose marginals are μ and ν. Formally this means that $\pi_{X\#}\gamma = \mu$ and $\pi_{Y\#}\gamma = \nu$, where $\pi_X : X \times Y \to X$ and $\pi_Y : X \times Y \to Y$ are the canonical projections. Denoting by $\Pi(\mu, \nu)$ the class of plannings, he wrote the following minimization problem

$$\min \left\{ \int_{X \times Y} c(x, y) \, d\gamma : \ \gamma \in \Pi(\mu, \nu) \right\}. \qquad (5)$$

Notice that $\Pi(\mu, \nu)$ is not empty, as the product $\mu \otimes \nu$ has μ and ν as marginals. Due to the convexity of the new constraint $\gamma \in \Pi(\mu, \nu)$ it turns out that weak topologies can be effectively used to provide existence of solutions to (5): this happens for instance whenever X and Y are Polish spaces and c is lower semicontinuous (see for instance [36]).

The connection between the Kantorovich formulation of the transport problem and Monge's original one can be seen noticing that any transport map t induces a planning γ, defined by $(Id \times t)_\#\mu$. This planning is concentrated on the graph of t in $X \times Y$ and it is easy to show that the converse holds, i.e. whenever γ is concentrated on a graph, then γ is induced by a transport map. Since any transport induces a planning with the same cost, it turns out that

$$\inf (3) \geq \min (5).$$

Moreover, by approximating any planning by plannings induced by transports, it can be shown that equality holds under fairly general assumptions (see for instance [3]). Therefore we can really consider the Kantorovich formulation of the transport problem as a weak formulation of the original problem.

The theory of disintegration of measures (see the Appendix) provides a very useful representation of plannings, and more generally of probability measures γ in $X \times Y$ whose first marginal is μ: there exist probability measures γ_x in Y such that $\gamma = \gamma_x \otimes \mu$, i.e.

$$\int_{X \times Y} \varphi(x,y) \, d\gamma(x,y) = \int_X \left(\int_Y \varphi(x,y) \, d\gamma_x(y) \right) d\mu(x).$$

for any bounded measurable function φ. In this sense we can consider a planning γ as a "stochastic" transport map $x \mapsto \gamma_x$, allowing the splitting of mass, and corresponding to a "deterministic" transport map only if γ_x is a Dirac mass for μ-a.e. $x \in X$. This representation of plannings also shows the close connection between the ideas of Kantorovich and of L.C. Young, who developed in the same years his theory of generalized controls (see [45, 46, 47]).

Kantorovich's weak solutions are by now considered as the "natural" solutions of the problem in Probability and in some related fields and, besides the general existence theorem mentioned above, general necessary and sufficient conditions for optimality, based on a duality formulation, have been found (see for instance [36], [24], [39], [40], [44]). Notice that, by the Choquet theorem, the linearity of the functional and the convexity of the constraint $\gamma \in \Pi(\mu, \nu)$ ensure that the minimum is achieved on an *extremal* point of $\Pi(\mu, \nu)$. Therefore, if extremal points were induced by transports one would get existence of transport maps directly from the Kantorovich formulation. It is not difficult to show that plannings γ induced by transports are extremal in $\Pi(\mu, \nu)$, since the disintegrated measures γ_x are Dirac masses; the converse holds in some particular cases, as

$$\mu = \sum_{i=1}^{N} \frac{1}{N} \delta_{x_i}, \qquad \nu = \sum_{i=1}^{N} \frac{1}{N} \delta_{y_i}$$

(by the well-known Birkhoff theorem) but unfortunately it is not true in general: for instance the measure $\gamma := \gamma_x \otimes \mathcal{L}^1 \llcorner [0,1]$ with $\gamma_x := \frac{1}{2}(\delta_x + \delta_{2-x})$ is not induced by a transport $y = t(x)$ but it is extremal in $\Pi(\mu, \nu)$ with $\nu := \frac{1}{2}\mathcal{L}^1 \llcorner [0,2]$, due to the fact that it is induced by the transport $x = t(y) = |y - 1|$. It turns out that the existence of optimal transport maps depends not only on the geometry of $\Pi(\mu, \nu)$, but also on the choice of the cost function c.

When $X = Y = \mathbb{R}^n$ and $c(x,y) = h(x - y)$ with h *strictly* convex and μ absolutely continuous with respect to $\mathcal{L}^n$, the duality methods yield that any optimal planning is induced by a transport; as a consequence, the optimal transport map exists and is unique (see [12], [13], [14], [24], [39], [40]).

The same results can be shown directly making a first variation in the dual formulation, bypassing the Kantorovich formulation (see [29], [15]). See also [34] for the extension of these results to a Riemannian setting.

The case when h is not strictly convex, corresponding to the original problem raised by Monge, is more subtle. Indeed, if $c(x,y) = |x - y|$ (the euclidean distance) then the standard duality methods provide information on the direction of transportation but no information on the distance $|x-y|$, at least when μ is absolutely continuous with respect to Lebesgue measure: to be precise one can show the existence of a 1-Lipschitz map $u : \mathbb{R}^n \to \mathbb{R}$ such that

$$(x,y) \in \operatorname{spt}\gamma \implies y \in \{x - t\nabla u(x) : t \geq 0\}$$

for μ-a.e. x. A similar result holds if the Euclidean norm is replaced by any strictly convex norm. Moreover, when $c(x,y) = \|x - y\|$ with $\|\cdot\|$ *not* strictly convex, then we have an even more dramatic loss of information about the location of y:

$$(x,y) \in \operatorname{spt}\gamma \implies y \in \{x - t(du(x))^* : t \geq 0\},$$

where, for L in the unit ball of $(\mathbb{R}^n)^*$ (the dual of $\mathbb{R}^n$), the set L^* consists of all vectors $v \in \mathbb{R}^n$ such that $\|v\| = 1$ and $L(v) = 1$.

The first attempt to bypass these difficulties came with the work of Sudakov [41], who claimed to have a solution for any distance cost function induced by a norm. Sudakov's approach is based on a clever decomposition of the space $\mathbb{R}^n$ in affine regions with variable dimension where the Kantorovich potential associated to the transport problem is an affine function. His strategy is to solve the transport problem in any of these regions, eventually getting an optimal transport map just by gluing all these transport maps. An essential ingredient in his proof is Proposition 78, where he states that, if $\mu \ll \mathcal{L}^n$, then the conditional measures induced by the decomposition are absolutely continuous with respect to the Lebesgue measure (of the correct dimension). However, it turns out that this property is not true in general even for the simplest decomposition, i.e. the decomposition in segments: G.Alberti, B.Kirchheim and D.Preiss [4] found an example of a compact faily of pairwise disjoint open segments in $\mathbb{R}^3$ such that the family M of their midpoints has strictly positive Lebesgue measure (the construction is a variant of previous examples due to A.S.Besicovitch and D.G.Larman [33]). In this case, choosing $\mu = \mathcal{L}^3 \llcorner M$, the conditional measures induced by the decomposition are Dirac masses. Therefore it is clear that this kind of counterexamples should be ruled out by some kind of additional "regularity" property of the decomposition. In this way the Sudakov strategy would be fully rigorous.

Several years later, Evans and Gangbo made a remarkable progress in [24], showing by differential methods the existence of a transport map, under the assumption that $\operatorname{spt}\mu \cap \operatorname{spt}\nu = \emptyset$, that the two measures are absolutely continuous with respect to $\mathcal{L}^n$ and that their densities are Lipschitz functions with compact support. The missing piece of information about the length of transportation is recovered by a p-laplacian approximation

$$-\text{div}\left(|\nabla u|^{p-2}\nabla u\right) = \mu - \nu, \qquad u \in H_0^1(B_R), \qquad R \gg 1$$

obtaining in the limit as $p \to +\infty$ a nonnegative function $a \in L^\infty(\mathbb{R}^n)$ and a 1-Lipschitz function u solving

$$-\text{div}\,(a\nabla u) = \mu - \nu, \qquad |\nabla u| = 1 \; \mathcal{L}^n\text{-a.e. on } \{a > 0\}.$$

The measure $\sigma := a\mathcal{L}^n$, the so-called *transport density*, plays an important rôle in the theory: its uniqueness and its regularity are discussed, under more general assumptions on μ and ν, in [27], [20], [21]. This measure appears in the *scalar mass optimization* problem studied in [9, 10, 11], and in [3] several equivalent representation of the transport density and its uniqueness have been studied.

Coming back to the transport problem with Euclidean distance (or, more generally, with a distance induced by a C^2 and uniformly convex norm), the first existence results for general absolutely continuous measures μ, ν with compact support have been independently obtained by L.Caffarelli, M.Feldman and R.Mc Cann in [16] and by N.Trudinger and L.Wang in [42]. Afterwards, the first author estabilished in [3] the existence of an optimal transport map assuming only that the initial measure μ is absolutely continuous, and the results of [16] and [42] have been extended to a Riemannian setting in [28]. All these proofs involve basically a Sudakov decomposition in transport rays, but the technical implementation of the idea is different from paper to paper: for instance in [16] a local change of variable is made, so that transport rays become parallel and Fubini theorem, in place of abstract disintegration theorems for measures, can be used. The proof in [3], instead, uses the co-area formula to show that absolute continuity with respect to Lebesgue measure is stable under disintegration.

In this paper we are particularly interested to the strategy pursued in [16], based on the approximation of the cost function $c(x,y) = \|x - y\|$ by the cost functions $c_\varepsilon(x,y) = \|x - y\|^{1+\varepsilon}$. The approximation is used in that paper to build a special Kantorovich potential u, by taking limits as $\varepsilon \downarrow 0$ of the potentials $(u_\varepsilon, u_\varepsilon^{c_\varepsilon})$ in the dual formulation (see Section 3 for a precise description of the dual formulation). The potential u is used to prescribe the geometry of transport rays and to build, by a 1-dimensional reduction, an optimal transport map. Here we give a new variational interpretation of the Caffarelli-Feldman-McCann approximation, based on the theory of asymptotic developments by Γ-convergence, developed by G.Anzellotti and S.Baldo in [7] (see also [8]). This new interpretation provides stronger results and, in particular, allows us to show that the family of optimal maps t_ε relative to the costs c_ε converges in measure as $\varepsilon \downarrow 0$ to the map built in [16]. However, since we don't assume a priori the existence of this map, our strategy provides at the same time an existence and a stability result for the Monge problem.

The underlying variational principle in our argument is that any limit of the optimal plannings γ_ε associated to t_ε is not only optimal for the Kantorovich problem, but optimal for the *secondary* variational problem

$$\min_{\gamma \in \Pi_1(\mu,\nu)} \int_{\mathbb{R}^n \times \mathbb{R}^n} \|x - y\| \ln(\|x - y\|) \, d\gamma, \tag{6}$$

where $\Pi_1(\mu, \nu)$ denotes the class of all optimal plannings for the Kantorovich problem (the entropy function in (6) comes from the Taylor expansion of c_ε around $\varepsilon = 0$, see also [11]). Then, we show that the secondary variational problem has a unique minimizer, and that this minimizer is induced by a transport map (a posteriori, the map built in [16]).

It is very likely, as the authors themselves of [16] suggest in the introduction of their paper, that the convergence of t_ε can still be proved working in the dual formulation, without appealing to our variational argument. However, we discovered that this new principle can be used in some situations to provide a "variational" decomposition in transport rays, bypassing the above mentioned difficulties in Sudakov's argument: in the forthcoming paper [5] we will show existence of optimal transport maps for distances induced by any "crystalline" norm $\| \cdot \|$ (whose unit ball is contained in finitely many hyperplanes and therefore not strictly convex) by looking at the approximation

$$c_\varepsilon(x,y) := \|x - y\| + \varepsilon|x - y| + \varepsilon^2|x - y|\ln(|x - y|).$$

Quite surprisingly, also in this situation we get full convergence as $\varepsilon \downarrow 0$ of t_ε to an optimal map t. Moreover, we will obtain existence of optimal transports for distances induced by *any* norm in the planar case $n = 2$.

We close this introduction by an analytic description of the content of this paper, conceived as a survey paper but also with original results, some of which are necessary for the forthcoming work [5].

In Section 3 we develop the duality theory for the Kantorovich problem. In view of the applications we have in mind (see Remark 7.1) we allow lower semicontinuous and possibly infinite cost functions, showing that also in this situation the c-monotonicity is a necessary condition for minimality (this is one of the key technical ingredients in [5]). We also discuss the problem of sufficiency of c-monotonocity: a general answer to this problem is not known, but we find a very general sufficient condition which seems not to be available in the literature (see Remark 3.1). We also provide a counterexample, but with a $+\infty$ valued cost function.

Section 4 contains the basic facts about the theory of Γ-asymptotic developments.

Section 5 reviews the theory of optimal transportation on the real line, for convex and nondecreasing cost functions. In this situation it is well known that, if μ has no atom, the unique nondecreasing map t pushing μ into ν is optimal, and it is the unique optimal map (even among plannings) when the cost function is nondecreasing and strictly convex. We show also a simple variant of this result (Theorem 5.2) where we drop the monotonicity assumption on the cost, to allow the entropy function as in (6).

Section 6 contains an abstract version of Sudakov's argument, based on a decomposition of the space in 1-dimensional transport rays, see Theorem 6.2.

We will apply this result to solve the transport problem in particular situations, see [5], [6]. In view of the counterexample [4], we make the assumption that the family of (maximal) rays is countably Lipschitz to ensure that any absolutely continuous measure μ with respect to $\mathcal{L}^n$ produces, after disintegration, a family of measures concentrated on 1-dimensional rays and absolutely continuous with respect to $\mathcal{H}^1$, therefore with no atom. As a consequence the 1-dimensional theory of the previous section applies to these measures. Notice also that, in view of the example in [33], the countable Lipschitz property seems to be necessary also to show that the family of estreme points of rays is Lebesgue negligible.

Section 7 contains the existence and stability result for Monge optimal transports mentioned above, under the same assumptions on the norm made in [16].

In Section 8 we show by a counterexample (or, rather, a class of counterexamples) that the absolute continuity assumption on μ is necessary. This is a distinctive feature of the linear case, since for strictly convex cost functions we have existence and uniqueness of optimal transport maps whenever μ has dimension strictly greater than $n - 1$ (see [30]).

Finally, the Appendix contains all basic facts about disintegration of measures needed in this paper. Of particular interest is Theorem 9.4, taken from [3], where we show stability of absolute continuity under disintegration, and the measurability criterion stated in Theorem 9.2.

2 Notation

In this section we fix our main notation and the terminology. We shall always assume that the measurable spaces we deal with are metric spaces endowed with the Borel σ-algebra, although this assumption could be easily weakened in some situations. Given a Borel map $f : X \to Y$, and given a positive and finite measure μ on X, we denote by $f_\#\mu$ its image, defined by $f_\#\mu(B) = \mu\left(f^{-1}(B)\right)$ for any Borel set $B \subset Y$. According to the change of variable formula we have

$$\int_Y \phi \, d\nu = \int_X \phi \circ f \, d\mu \qquad \text{for any bounded Borel function } \phi : Y \to \mathbb{R}.$$

We denote by spt μ the support of μ, i.e. the closed set of all points $x \in X$ such that $\mu(B_r(x)) > 0$ for any $r > 0$. We say that μ is *concentrated* on a Borel set B if $\mu(X \setminus B) = 0$. If X is separable then μ is concentrated on spt μ and the support is the minimal closed set on which μ is concentrated. On the other hand, a measure can be concentrated on a set much smaller than the support: for instance the probability measure

$$\mu := \sum_{n=0}^{\infty} 2^{-1-n} \delta_{q_n},$$

where $\{q_n\}$ is an enumeration of the rational numbers, has $\mathbb{R}$ as support but it is concentrated on $\mathbf{Q}$.

In the following table we resume the notation used without further explaination throughout the text:

$\mathcal{L}^n$	Lebesgue measure in $\mathbb{R}^n$
$\mathcal{H}^k$	Hausdorff k-dimensional measure in $\mathbb{R}^n$
$\mathbf{S}^{n-1}$	unit sphere in $\mathbb{R}^n$
$\mathcal{B}(X)$	Borel σ-algebra of X
$\mathrm{Lip}(X)$	real valued Lipschitz functions defined on X
$\mathrm{Lip}_1(X)$	functions in $\mathrm{Lip}(X)$ with Lipschitz constant not greater than 1
π_0, π_X	projection $X \times Y \ni (x,y) \mapsto x \in X$
π_1, π_Y	projection $X \times Y \ni (x,y) \mapsto y \in Y$
$\mathcal{S}_o(X)$	open oriented segments $]x,y[$ with $x, y \in X$, $x \neq y$
$\mathcal{S}_c(X)$	closed oriented segments $[x,y]$ with $x, y \in X$, $x \neq y$
$\mathcal{M}_+(X)$	positive and finite Radon measures in X
$\mathcal{P}(X)$	probability measures in X
$\mu \llcorner B$	restriction of μ to B, defined by $\chi_B \mu$.

3 Duality and optimality conditions

In this section we look for general necessary and sufficient optimality conditions for the Kantorovich problem (5). We make fairly standard assumptions on the spaces X, Y, assuming them to be locally compact and separable (these assumptions can be easily relaxed, see for instance [36]), but we look for general lower semicontinuous cost functions $c : X \times Y \to [0, +\infty]$, allowing in some cases the value $+\infty$. This extension is important in view of the applications we have in mind (see Remark 7.1 and [5]). See also [36] for more general versions of the duality formula.

Theorem 3.1 (Duality formula). *The minimum of the Kantorovich problem is equal to*

$$\sup \left\{ \int_X \varphi(x) \, d\mu(x) + \int_Y \psi(y) \, d\nu(y) \right\} \tag{7}$$

where the supremum runs among all pairs $(\varphi, \psi) \in L^1(X, \mu) \times L^1(Y, \nu)$ such that $\varphi(x) + \psi(y) \leq c(x,y)$.

Proof. This identity is well-known if c is bounded, see for instance [36], [44]. In the general case it suffices to approximate c from below by an increasing sequence of bounded continuous functions c_h, defined for instance by

$$c_h(x,y) := \min \left\{ c(x',y') \wedge h + h d_X(x,x') + h d_Y(y,y') \right\},$$

noticing that a simple compactness argument gives

$$\min\left\{\int_{X\times Y} c_h\, d\gamma : \gamma \in \Pi(\mu,\nu)\right\} \quad \uparrow \quad \min\left\{\int_{X\times Y} c\, d\gamma : \gamma \in \Pi(\mu,\nu)\right\}$$

and that any pair (φ,ψ) such that $\varphi + \psi \leq c_h$ is admissible in (7).

We recall briefly the definitions of c-transform, c-concavity and c-cyclical monotonicity, referring to the papers [22], [37], [30] and to the book [36] for a more detailed analysis.

For $u : X \to \overline{\mathbf{R}}$, the *c-transform* $u^c : Y \to \overline{\mathbf{R}}$ is defined by

$$u^c(y) := \inf_{x\in X} c(x,y) - u(x)$$

with the convention that the sum is $+\infty$ whenever $c(x,y) = +\infty$ and $u(x) = +\infty$. Analogously, for $v : Y \to \overline{\mathbf{R}}$, the *c-transform* $v^c : X \to \overline{\mathbf{R}}$ is defined by

$$v^c(x) := \inf_{y\in Y} c(x,y) - v(y)$$

with the same convention when an indetermination of the sum is present.

We say that $u : X \to \overline{\mathbf{R}}$ is *c-concave* if $u = v^c$ for some v; equivalently, u is c-concave if there is some family $\{(y_i, t_i)\}_{i\in I} \subset Y \times \overline{\mathbf{R}}$ such that

$$u(x) = \inf_{i\in I} c(x,y_i) + t_i \qquad \forall x \in X.$$

An analogous definition can be given for functions $v : Y \to \overline{\mathbf{R}}$.

It is not hard to show that $u^{cc} \geq u$ and that equality holds if and only if u is c-concave. Analogously, $v^{cc} \geq v$ and equality holds if and only if v is c-concave.

Finally, we say that $\Gamma \subset X \times Y$ is *c-monotone* if

$$\sum_{i=1}^{n} c(x_i, y_{\sigma(i)}) \geq \sum_{i=1}^{n} c(x_i, y_i)$$

whenever $(x_1, y_1), \ldots, (x_n, y_n) \in \Gamma$ and σ is a permutation of $\{1, \ldots, n\}$.

Theorem 3.2 (Necessary and sufficient optimality conditions).
(Necessity) If $\gamma \in \Pi(\mu,\nu)$ is optimal and $\int_{X\times Y} c\, d\gamma < +\infty$, then γ is concentrated on a c-monotone Borel subset of $X \times Y$.
(Sufficiency) Assume that c is real-valued, γ is concentrated on a c-monotone Borel subset of $X \times Y$ and

$$\mu\left(\left\{x \in X : \int_Y c(x,y)\, d\nu(y) < +\infty\right\}\right) > 0, \tag{8}$$

$$\nu\left(\left\{y \in Y : \int_X c(x,y)\, d\mu(x) < +\infty\right\}\right) > 0. \tag{9}$$

Then γ is optimal, $\int_{X\times Y} c\, d\gamma < +\infty$ and there exists a maximizing pair (φ,ψ) in (7) with φ c-concave and $\psi = \varphi^c$.

Proof. Let (φ_n, ψ_n) be a maximizing sequence in (7) and let $c_n = c - \varphi_n - \psi_n$. Since

$$\int_{X \times Y} c_n \, d\gamma = \int_{X \times Y} c \, d\gamma - \int_X \varphi_n \, d\mu - \int_Y \psi_n \, d\nu \to 0$$

and $c_n \geq 0$ we can find a subsequence $c_{n(k)}$ and a Borel set Γ on which γ is concentrated and c is finite, such that $c_{n(k)} \to 0$ on Γ. If $\{(x_i, y_i)\}_{1 \leq i \leq p} \subset \Gamma$ and σ is a permutation of $\{1, \ldots, p\}$ we get

$$\sum_{i=1}^p c(x_i, y_{\sigma(i)}) \geq \sum_{i=1}^p \varphi_{n(k)}(x_i) + \psi_{n(k)}(y_{\sigma(i)})$$

$$= \sum_{i=1}^p \varphi_{n(k)}(x_i) + \psi_{n(k)}(y_i) = \sum_{i=1}^p c(x_i, y_i) - c_{n(k)}(x_i, y_i)$$

for any k. Letting $k \to \infty$ the c-monotonicity of Γ follows.

Now we show the converse implication, assuming that (8) and (9) hold. We denote by Γ a Borel and c-monotone set on which γ is concentrated; without loss of generality we can assume that $\Gamma = \cup_k \Gamma_k$ with Γ_k compact and $c|_{\Gamma_k}$ continuous. We choose continuous functions c_l such that $c_l \uparrow c$ and split the proof in several steps.

Step 1. There exists a c-concave Borel function $\varphi : X \to [-\infty, +\infty)$ such that $\varphi(x) > -\infty$ for μ-a.e. $x \in X$ and

$$\varphi(x') \leq \varphi(x) + c(x', y) - c(x, y) \qquad \forall x' \in X, \ (x, y) \in \Gamma. \tag{10}$$

To this aim, we use the explicit construction given in the generalized Rockafellar theorem in [37], setting

$$\varphi(x) := \inf\{c(x, y_p) - c(x_p, y_p) + c(x_p, y_{p-1}) - c(x_{p-1}, y_{p-1}) \\ + \cdots + c(x_1, y_0) - c(x_0, y_0)\}$$

where $(x_0, y_0) \in \Gamma_1$ is fixed and the infimum runs among all integers p and collections $\{(x_i, y_i)\}_{1 \leq i \leq p} \subset \Gamma$.

It can be easily checked that

$$\varphi = \lim_{p \to \infty} \lim_{m \to \infty} \lim_{l \to \infty} \varphi_{p,m,l},$$

where

$$\varphi_{p,m,l}(x) := \inf\{c_l(x, y_p) - c(x_p, y_p) + c_l(x_p, y_{p-1}) - c(x_{p-1}, y_{p-1}) \\ + \cdots + c_l(x_1, y_0) - c(x_0, y_0)\}$$

and the infimum is made among all collections $\{(x_i, y_i)\}_{1 \leq i \leq p} \subset \Gamma_m$. As all functions $\varphi_{p,m,l}$ are upper semicontinuous we obtain that φ is a Borel function.

Arguing as in [37] it is straightforward to check that $\varphi(x_0) = 0$ and (10) holds. Choosing $x' = x_0$ we obtain that $\varphi > -\infty$ on $\pi_X(\Gamma)$ (here we use the assumption that c is real-valued). But since γ is concentrated on Γ the Borel set $\pi_X(\Gamma)$ has full measure with respect to $\mu = \pi_{X\#}\gamma$, hence $\varphi \in \mathbb{R}$ μ-a.e.

Step 2. Now we show that $\psi := \varphi^c$ is ν-measurable, real-valued ν-a.e. and that

$$\varphi + \psi = c \quad \text{on } \Gamma. \tag{11}$$

It suffices to study ψ on $\pi_Y(\Gamma)$: indeed, as γ is concentrated on Γ, the Borel set $\pi_Y(\Gamma)$ has full measure with respect to $\nu = \pi_{Y\#}\gamma$. For $y \in \pi_Y(\Gamma)$ we notice that (10) gives

$$\psi(y) = c(x,y) - \varphi(x) \in \mathbb{R} \qquad \forall x \in \Gamma_y := \{x : (x,y) \in \Gamma\}.$$

In order to show that ψ is ν-measurable we use the disintegration $\gamma = \gamma_y \otimes \nu$ of γ with respect to y (see the appendix) and notice that the probability measure γ_y is concentrated on Γ_y for ν-a.e. y, therefore

$$\psi(y) = \int_X c(x,y) - \varphi(x)\, d\gamma_y(x) \qquad \text{for } \nu\text{-a.e. } y.$$

Since $y \mapsto \gamma_y$ is a Borel measure-valued map we obtain that ψ is ν-measurable.

Step 3. We show that φ^+ and ψ^+ are integrable with respect to μ and ν respectively (here we use (8) and (9)). By (8) we can choose x in such a way that $\int_Y c(x,y)\, d\nu(y)$ is finite and $\varphi(x) \in \mathbb{R}$, so that by integrating on Y the inequality $\psi^+ \le c(x,\cdot) + \varphi^-(x)$ we obtain that $\psi^+ \in L^1(Y,\nu)$. The argument for φ^+ uses (9) and is similar.

Step 4. Conclusion. The semi-integrability of φ and ψ gives the null-lagrangian identity

$$\int_{X \times Y} (\varphi + \psi)\, d\tilde\gamma = \int_X \varphi\, d\mu + \int_Y \psi\, d\nu \in \mathbb{R} \cup \{-\infty\} \qquad \forall \tilde\gamma \in \Pi(\mu,\nu),$$

so that choosing $\tilde\gamma = \gamma$ we obtain from (11) that $\int_{X \times Y} c\, d\gamma < +\infty$ and $\varphi \in L^1(X,\mu)$, $\psi \in L^1(Y,\nu)$. Moreover, for any $\tilde\gamma \in \Pi(\mu,\nu)$ we get

$$\int_{X \times Y} c\, d\tilde\gamma \ge \int_{X \times Y} (\varphi + \psi)\, d\tilde\gamma = \int_X \varphi\, d\mu + \int_Y \psi\, d\nu$$

$$= \int_{X \times Y} (\varphi + \psi)\, d\gamma = \int_\Gamma (\varphi + \psi)\, d\gamma = \int_{X \times Y} c\, d\gamma.$$

This chain of inequalities gives that γ is optimal and, at the same time, that (φ, ψ) is optimal in (7).

We say that a Borel function $\varphi \in L^1(X,\mu)$ is a *maximal Kantorovich potential* if (φ, φ^c) is a maximizing pair in (7). In many applications it is useful to write the optimality conditions using a maximal Kantorovich potential, instead of the cyclical monotonicity.

Theorem 3.3. *Let $\mu \in \mathcal{P}(X)$, $\nu \in \mathcal{P}(Y)$, assume that (8) and (9) hold, that c is real-valued and the sup in (7) is finite. Then there exists a maximizing pair (φ, φ^c) in (7) and $\gamma \in \Pi(\mu, \nu)$ is optimal if and only if*

$$\varphi(x) + \varphi^c(y) = c(x, y) \qquad \gamma\text{-a.e. in } X \times Y.$$

Proof. The existence of a maximizing pair is a direct consequence of the sufficiency part of the previous theorem, choosing an optimal γ and (by the necessity part of the statement) a c-monotone set on which γ is concentrated.

If γ is optimal then

$$\int_{X \times Y} (c - \varphi - \varphi^c)\, d\gamma = \int_{X \times Y} c\, d\gamma - \int_X \varphi\, d\mu - \int_Y \varphi^c\, d\nu = 0.$$

As the integrand is nonnegative, it must vanish γ-a.e. The converse implication is analogous.

Remark 3.1. The assumptions (8), (9) are implied by

$$\int_{X \times Y} c(x, y)\, d\mu \otimes \nu(x, y) < +\infty. \tag{12}$$

In turn, (12) is weaker than the condition

$$c(x, y) \le a(x) + b(y) \quad \text{with} \quad a \in L^1(\mu),\ b \in L^1(\nu)$$

considered in Theorem 2.3.12 of Part I of [36].

The following example shows that some kind of finiteness/integrability condition seems to be necessary in order to infer minimality from cyclical monotonicity. It is interesting to notice that in very specific cases (but important for the applications) like $X = Y = \mathbb{R}^n$ and $c(x, y) = |x - y|^2$ it is not presently clear whether cyclical monotonicity implies minimality *without* additional conditions, e.g. the finiteness of the moments $\int |x|^2\, d\mu$, $\int |y|^2\, d\nu$ (see the Open problem 16 in Chapter 3 of [44]).

Example 3.1. Given $\alpha \in [0, 1] \setminus \mathbf{Q}$, let $\varphi : [0, 1] \times [0, 1] \to [0, +\infty]$ be defined as follows:

$$\varphi(x, y) := \begin{cases} 1 & \text{if } y = x \\ 2 & \text{if } y = x \oplus \alpha \\ +\infty & \text{otherwise,} \end{cases}$$

where $\oplus : [0, 1] \times [0, 1] \to [0, 1]$ is the sum modulo 1: Figure 1 shows this function. Let us then consider the transport problem in $\Omega = [0, 2] \subset \mathbb{R}$ with $\mu = \mathcal{L}^1 \llcorner [0, 1]$, $\nu = \mathcal{L}^1 \llcorner [1, 2]$ and with any cost c such that $c(x, y) = \varphi(x, y - 1)$ whenever $0 \le x \le 1 \le y \le 2$. Clearly the unique optimal plan of transport is

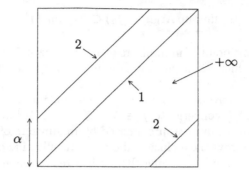

Fig. 1. The function φ in Example 3.1

$\bar{\gamma} = (Id, 1 + Id)_{\#} \mathcal{L}^1 \llcorner [0,1]$, while we will prove that also the support of the non-optimal plan $\gamma = \left(Id, 1 + (Id \oplus \alpha)\right)_{\#} \mathcal{L}^1 \llcorner [0,1]$ is c-monotone.

If not, there would be a minimal set of couples $(x_1, y_1), (x_2, y_2), \dots,$ (x_n, y_n) in the support of γ with the property that

$$d(x_1, y_1) + \cdots + d(x_n, y_n) > d(x_2, y_1) + d(x_3, y_2) + \cdots + d(x_1, y_n);$$

but $(x_i, y_i) \in \operatorname{spt}\gamma$ means $y_i = 1 + (x_i \oplus \alpha)$: moreover, since the preceding inequality assures $d(x_{i+1}, y_i) < +\infty$, we can infer that $y_i = 1 + x_{i+1}$ or $y_i = 1 + (x_{i+1} \oplus \alpha)$, and then $x_{i+1} = x_i \oplus \alpha$ or $x_{i+1} = x_i$. Since the second possibility is incompatible with the minimality of the set, it must be $x_{i+1} = x_i \oplus \alpha$; applying this equality n times, we find $x_1 = x_1 \oplus n\alpha$, which is impossible since α is not a rational number.

4 Γ-convergence and Γ-asymptotic expansions

In this section we recall some basic facts about Γ-convergence and we present the essential aspects of the theory of Γ-asymptotic expansions, first introduced in [7] (see also [8] for an application of this theory in elasticity). A general reference for the theory of Γ-convergence is [18].

Let X be a compact metric space and let us denote by $S_-(X)$ the collection of all lower semicontinuous functions $f : X \to \overline{\mathbf{R}}$. We say that a sequence $(f_h) \subset S_-(X)$ Γ-converges to $f \in S_-(X)$ if the following two properties hold for any $x \in X$:

(a) $\liminf f_h(x_h) \geq f(x)$ for any sequence $x_h \to x$;

(b) there exists a sequence $x_h \to x$ such that $f_h(x_h) \to f(x)$.

It can be shown that the Γ-convergence is induced by a metric d_Γ and that $(S_-(X), d_\Gamma)$ is a compact metric space.

If (f_h) Γ-converges to f then $m_h := \min_X f_h \to m := \min_X f$ and, in addition,

$$\limsup_{h\to\infty} \text{Argmin}\,(f_h) \subset \text{Argmin}\,(f). \tag{13}$$

In words, any limit point of minimizers of f_h minimizes f. The same is true for sequences (x_h) which are asymptotically minimizing, i.e. such that $f_h(x_h) - m_h \to 0$.

However, the converse is not true in general: for instance the functions $f_h(x) = 1/h \wedge |x|$ Γ-converge to $f \equiv 0$ in $X = [-1,1]$ but $x = 0$ is the only minimizer of f that can be approximated by minimizers of f_h.

In order to improve the inclusion above, G.Anzellotti and S.Baldo proposed the following procedure: assuming that m is a real number, they proposed to consider the new functions

$$f_h' := \frac{f_h - m}{\varepsilon_h}$$

for suitable positive infinitesimals ε_h. Assuming that f_h' Γ-converge to f' (this is not really restrictive, by the compactness of $S_-(X)$), the following result holds:

Theorem 4.1. *The functional f' is equal $+\infty$ out of Argmin f, hence Argmin $f' \subset$ Argmin f. Moreover*

$$\limsup_{h\to\infty} \text{Argmin}\,(f_h) \subset \text{Argmin}\,(f'). \tag{14}$$

Proof. If $f'(x) < +\infty$ there exists a sequence (x_h) converging to x, by condition (b), such that $f_h'(x_h)$ is bounded above. As

$$f_h(x_h) = m + \varepsilon_h f_h'(x_h) \le m_h + o(1)$$

the sequence (x_h) is asymptotically minimizing and therefore $x \in$ Argmin f. Finally (14) is a direct consequence of (13), noticing that Argmin $f_h =$ Argmin f_h'.

If ϵ_h have been properly chosen, so that $m' := \min_X f' \in \mathbb{R}$, then the convergence of $\min_X f_h' = (m_h - m)/\varepsilon_h$ to m' gives the expansion

$$m_h = m + m'\varepsilon_h + o(\varepsilon_h).$$

This procedure can of course be iterated, giving further restrictions on the set of limit points of minimizers of f_h and higher order expansions of the difference $m_h - m$.

5 1-dimensional theory

In this section we recall some aspects of the theory of optimal transportation in $\mathbb{R}$. In this case, at least when μ has no atom, there is a canonical transport map obtained by monotone rearrangement. This map is optimal whenever the cost is a nondecreasing and convex function of the distance.

Theorem 5.1. *Let μ, $\nu \in \mathcal{P}(\mathbb{R})$, μ without atoms, and let*

$$G(x) := \mu\left((-\infty, x)\right), \qquad F(y) := \nu\left((-\infty, y)\right)$$

be respectively the distribution functions of μ, ν. Then

(i) The nondecreasing function $t : \mathbb{R} \to \overline{\mathbb{R}}$ defined by

$$t(x) := \sup\{y \in \mathbb{R} : F(y) \le G(x)\}$$

(with the convention $\sup \emptyset = -\infty$) maps μ into ν. Any other nondecreasing map t' such that $t'_\# \mu = \nu$ coincides with t on $\mathrm{spt}\,\mu$ up to a countable set.

(ii) If $\phi : [0, +\infty) \to \mathbb{R}$ is nondecreasing and convex, then t is an optimal transport between μ and ν relative to the cost $c(x, y) = \phi(|x-y|)$. Moreover t is the unique optimal transport map if ϕ is strictly convex.

For the proof the reader may consult [3], [30], [44]).

Notice that the monotonicity constraint forces us to take $\overline{\mathbb{R}}$ as range of t (for instance this happens when μ has compact support and $\mathrm{spt}\,\nu = \mathbb{R}$). However, since $t_\# \mu = \nu$, the half-lines $\{t = \pm\infty\}$ are μ-negligible.

In view of the applications we have in mind (where $\phi(t)$ could be $t \ln t$) we are interested in dropping the assumption that ϕ is nondecreasing. This can be done under a suitable compatibility condition between μ and ν, expressed in (15) below, by restricting the class of competitors γ.

Theorem 5.2. *Let μ, $\nu \in \mathcal{P}(\mathbb{R})$, μ without atoms and let t be the map in Theorem 5.1. Assume that μ and ν have finite first order moments and*

$$\mathcal{A} := \{\gamma \in \Pi(\mu, \nu) : \mathrm{spt}\,\gamma \subset \{(x, y) : y \ge x\}\} \ne \emptyset. \tag{15}$$

Then $t(x) \ge x$ for μ-a.e. $x \in \mathbb{R}$ and $\gamma_t = (Id \times t)_\# \mu$ is a solution of the problem

$$\min_{\gamma \in \mathcal{A}} \int_{\mathbb{R} \times \mathbb{R}} \phi(|y - x|)\, d\gamma \tag{16}$$

whenever $\phi : [0, +\infty) \to \mathbb{R}$ is a convex function bounded from below. If ϕ is strictly convex and the minimum in (16) is finite, then γ_t is the unique solution.

Proof. We argue as in [30] and we consider the strictly convex case only (the general case follows by a simple perturbation argument). Since $\mathcal{A}$ is weakly compact, we can find an optimal $\gamma \in \mathcal{A}$ for (16). Assuming that γ has finite energy, by the construction in [30] one can show that $\Gamma := \mathrm{spt}\,\gamma$ satisfies the restricted monotonicity condition

$$\phi(y_1 - x_1) + \phi(y_2 - x_2) \le \phi(y_2 - x_1) + \phi(y_1 - x_2)$$

whenever (x_1, y_1), $(x_2, y_2) \in \Gamma$ and $x_1 < y_2$, $x_2 < y_1$ (indeed, in this case the additional constraint that competitors must be in $\mathcal{A}$ is not effective).

Now we show the implication:

$$(x_1, y_1) \in \Gamma, \quad (x_2, y_2) \in \Gamma, \quad x_1 < x_2 \quad \Longrightarrow \quad y_1 \leq y_2. \qquad (17)$$

Assuming by contradiction that $y_1 > y_2$, since $x_i \leq y_i$, $i = 1, 2$, we obtain $x_1 < x_2 \leq y_2 < y_1$. In this case, setting $a = x_2 - x_1$, $b = y_2 - x_2$, $c = y_1 - y_2$, the cyclical monotonicity of Γ gives

$$\phi(a + b + c) + \phi(b) \leq \phi(a + b) + \phi(b + c).$$

On the other hand, since $c > 0$ the strict convexity of ϕ gives

$$\phi(a + b + c) - \phi(b + c) > \phi(a + b) - \phi(b)$$

and therefore a contradiction.

By (17) we obtain that the vertical sections Γ_x of Γ are ordered, i.e. $y_1 \in \Gamma_{x_1} \leq y_2 \in \Gamma_{x_2}$ whenever $x_1 < x_2$. As a consequence the set of all x such that Γ_x is not a singleton is at most countable (since we can index with this set a family of pairwise disjoint open intervals), and therefore μ-negligible. As $\mathrm{spt}\, \gamma_x \subset \Gamma_x$, it follows that γ_x is a Dirac mass for μ-a.e. x. Setting $\gamma_x = \delta_{t'(x)}$ the map t' is nondecreasing by (17), satisfies $t'(x) \geq x$ because $\gamma \in \mathcal{A}$ and maps μ into ν because $\gamma = (Id \times t')_{\#}\mu$. By Theorem 5.1(i) we obtain $t' = t$ μ-a.e.

The proof is finished by showing that problem (16) is non trivial (i.e. the minimum is finite) for at least one stricly convex ϕ. This follows by the assumption on the finiteness of first moments, choosing $\phi(t) := \sqrt{1 + t^2} - 1$.

6 Transport rays and transport set

Given $\Gamma \subset \mathbf{R}^n \times \mathbf{R}^n$, we define the notions of *transport ray*, *ray direction*, *transport set*, *fixed point*, *maximal transport ray*. The terminology is close to the one adopted in [24] and [16], with the difference that our definitions depend on Γ rather than a Kantorovich potential. We reconcile with the other approaches in (24) and in Theorem 6.2.

(Transport ray) We say that $]x, y[$ is a *transport ray* if $x \neq y$ and $(x, y) \in \Gamma$.
(Ray direction) Given a transport ray $]x, y[$, we denote its direction by

$$\tau(x, y) := \frac{(y - x)}{|y - x|}.$$

(Transport sets) We denote by T_Γ the union of all transport rays relative to Γ, i.e.

$$\bigcup_{(x,y) \in \Gamma}]x, y[.$$

We define also T_Γ^l as the union of all sets $[x, y[$, as $(x, y) \in \Gamma$, and T_Γ^r as the union of all sets $]x, y]$, as $(x, y) \in \Gamma$ (by convention $[x, y[=]x, y] = \{x\}$ if $x = y$).

(Fixed points) We say that x is a fixed point if $(x, x) \in \Gamma$ and $(x, y) \notin \Gamma$ for any $y \neq x$. We denote by F_Γ the set of all fixed points.

(Maximal transport ray) We say that an open interval $S \subset \mathbf{R}^n$ (possibly unbounded) is a maximal transport ray if

(a) for any $z \in S$ there exists $(x, y) \in \Gamma$ with $z \in]x, y[$;

(b) any open interval containing S and satisfying (a) coincides with S.

Notice that maximal transport rays need not be transport rays: for instance if $\Gamma = \{(1/(k + 2), 1/k)\}$, for $k \geq 1$ integer, then $(0, 1)$ is a maximal transport ray but not a transport ray.

In the following we shall always assume that Γ is a σ-compact set. This ensures that all sets T_Γ, T_Γ^l, T_Γ^r, F_Γ associated to Γ are Borel sets. Notice also that

$$F_\Gamma = \pi_0(\Gamma \cap \Delta) \setminus \pi_0(\Gamma \setminus \Delta),$$

where Δ is the diagonal of $\mathbb{R}^n \times \mathbb{R}^n$.

Clearly any transport ray is contained in a unique maximal transport ray. As a consequence, any point in T_Γ is contained in a maximal transport ray. Under the no-crossing condition

$$[x, y] \cap [x', y'] \neq \emptyset \implies x = x' \text{ or } y = y' \qquad \text{whenever } \tau(x, y) \neq \tau(x', y') \tag{18}$$

(meaning that two closed rays with different orientations can meet only at a common endpoint) it is also immediate to check that this maximal transport ray is unique. Therefore Γ induces a map $\pi_r : T_\Gamma \to S_o(\mathbb{R}^n)$ which associates to any point the maximal transport ray containing it. We also denote by $\tau_r : T_\Gamma \to \mathbf{S}^{n-1}$ the map which gives the direction of the transport ray.

Since we used the open segments to define the transport set and the maximal transport ray, we need to take into account also the extreme points (which are not in T_Γ) of the maximal transport rays.

The following proposition shows that only points in $T_\Gamma^l \cup T_\Gamma^r$ can carry some mass, and therefore are relevant for the transport problem.

Proposition 6.1. *(i) Any point in $T_\Gamma^l \setminus (T_\Gamma \cup F_\Gamma)$ (respectively $T_\Gamma^r \setminus (T_\Gamma \cup F_\Gamma)$) is a left (resp. right) extreme point of a maximal transport ray.*

(ii) If $\gamma \in \Pi(\mu, \nu)$ is concentrated on Γ, then μ is concentrated on T_Γ^l and ν is concentrated on T_Γ^r.

Proof. (i) If $x \in T_\Gamma^l \setminus F_\Gamma$ there exists $y \neq x$ such that $(x, y) \in \Gamma$. If $x \notin T_\Gamma$ the maximal transport ray containing $]x, y[$ must have x as left extreme point.

(ii) Clearly μ is concentrated on L, the projection of Γ on the first factor. If $x \in L$ there exists $y \in \mathbb{R}^n$ such that $(x, y) \in \Gamma$, and therefore $x \in T_\Gamma^l$. The argument for ν is similar.

The converse implication in Proposition 6.1(i) does not hold, as shown by the previous example of a maximal transport ray which is not a transport ray.

Definition 6.1 (Metric on $S_c(\mathbb{R}^n)$ and $S_o(\mathbb{R}^n)$). In the following we need to define a metric structure (actually we would need only a measurable one) on the spaces $S_c(\mathbb{R}^n)$ and $S_o(\mathbb{R}^n)$. When one considers only bounded oriented segments the natural metric comes from the embedding into $\mathbb{R}^n \times \mathbb{R}^n$, by looking at the distances between the two left extreme points and the two right extreme points. In our case, since we allow halflines and lines as segments, we define the metric in $S_c(\mathbb{R}^n)$ as

$$d(S, S') := \sum_{R=1}^{\infty} 2^{-R} \frac{|x_R - x'_R| + |y_R - y'_R|}{1 + |x_R - x'_R| + |y_R - y'_R|},$$

where $[x_R, y_R]$ (respectively $[x'_R, y'_R]$) is the intersection of S (resp. S') with the closed ball $\overline{B}_R$. Since any open segment is in one to one correspondence with a closed segment (remember that singletons do not belong to $S_c(\mathbb{R}^n)$) we define a metric in $S_o(\mathbb{R}^n)$ in such a way that this correspondence is an isometry.

It is not hard to check that $S_o(\mathbb{R}^n)$ and $S_c(\mathbb{R}^n)$ are locally compact and separable metric spaces.

Now we prove a measurability result about the map π_r.

Lemma 6.1. *If Γ is σ-compact and (18) holds, then the map π_r which associates to any point in T_Γ the maximal transport ray in $S_o(\mathbb{R}^n)$ containing it is Borel.*

Proof. Since Γ is σ-compact, let us write $\Gamma = \bigcup_{i \in \mathbb{N}} \Gamma_i$, where each Γ_i is compact; we define also $\varphi : P(S_o) \to P(\mathbb{R}^n)$, denoting by $P(X)$ the subsets of X, as follows:

$$\varphi(C) := \{x \in \mathbb{R}^n : \exists S \in C \, s.t. \, x \in S\} \qquad \forall C \subset S_o.$$

The first thing we can note is the following
Claim: If $C \subset S_o(\mathbb{R}^n)$ is closed, then $\varphi(C) \subset \mathbb{R}^n$ is Borel.
Let us first of all consider the case when C is compact: the assert follows directly, recalling Definition 6.1: if we would work with the closed segments, i.e. with $S_c(\mathbb{R}^n)$ instead of $S_o(\mathbb{R}^n)$, then the image of C were easily seen to be compact, and then $\varphi(C)$ is a countable union of compact sets, and then a Borel set. In general, if C is closed, it is a countable union of compact sets, and then $\varphi(C)$ is a countable union of Borel sets, thus Borel.

To prove the thesis, given a compact $C \subset S_o(\mathbb{R}^n)$, it suffices to show that $\pi_r^{-1}(C)$ is a Borel set: to do this, first of all we define

$$C_i := \left\{ \,]sx + (1-s)y, ty + (1-t)x[\quad s.t. \,]x, y[\in C, \, 0 \le s, t \le \frac{1}{2^i} \right\},$$

which is easily seen to be closed (in fact, it is compact). Given any $p > 0$, we will moreover denote by G_p the (closed) set of all the open segments in $\mathbb{R}^n$ of lenght p. We need to define now the sets $\Gamma_{i,j} \subset S_o(\mathbb{R}^n)$ with $i, j \in \mathbb{N}$; to begin, for any $X \subset S_o(\mathbb{R}^n)$ we will denote by $Sub(X)$ the set of all its open subsegments: in other words, $]x, y[\in Sub(X)$ if and only if there exists $]z, w[\in X$ such that $]x, y[\subset]z, w[$. For $j = 1$, then, we fix $\Gamma_{i,1} := Sub(\Gamma_i)$, while for $j = 2$, it will be

$$\Gamma_{i,2} := Sub\left(\Gamma_{i,1} \bigcup \{]x, y[=]x, w[\cup]z, y[\ s.t. \ ]x, w[,]z, y[\in \Gamma_{i,1}, |z - w| \geq 1\}\right).$$

In words, we add to $\Gamma_{i,1}$ some open segments which are union of two segments in $\Gamma_{i,1}$, and then consider again all the possible subsegments; it is easy to note that $\Gamma_{i,2}$ is closed thanks to the request $|z - w| \geq 1$ in the last definition (in fact, we made that assumption only to ensure the closedness of $\Gamma_{i,2}$). In general, we will write

$$\Gamma_{i,j+1} := Sub(\Gamma_{i,j} \cup \{]x, y[=]x, w[\cup]z, y[\ s.t. \ ]x, w[,]z, y[\in \Gamma_{i,j}, |z - w| \geq 1/j\})$$

which generalizes the definition of $\Gamma_{i,2}$. One can note that $\Gamma_{i,i}$ is an increasing sequence of subsets of $S_o(\mathbb{R}^n)$ and that

$$\Gamma_{i,i} \longrightarrow Sub\left(\pi_r(T_\Gamma)\right) \qquad for \ i \to +\infty.$$

The last step to reach the thesis is then to note that

$$\pi_r{}^{-1}(C) = \bigcap_{i \in \mathbb{N}} \bigcup_{m \in \mathbb{N}} \bigcup_{p \in \mathbb{Q}^+} \bigcap_{n \geq m} \left((\varphi(\Gamma_n \cap C_i \cap G_p)) \setminus (\varphi(\Gamma_n \cap G_{p+1/2^{i-1}}))\right),$$

which assures $\pi_r{}^{-1}(C)$ to be Borel. To be convinced of the last equation, let us restrict our attention to the case when Γ is associated to a single maximal transport ray $]x, y[$ of lenght l: then we can note that, for n sufficiently large,

$$\left(\varphi(\Gamma_n \cap G_p)\right) \setminus \left(\varphi(\Gamma_n \cap G_{p+1/2^{i-1}})\right)$$

is empty unless $p \leq l \leq p + 1/2^{i-1}$. Thus

$$\bigcup_{m \in \mathbb{N}} \bigcup_{p \in \mathbb{Q}^+} \bigcap_{n \geq m} \left((\varphi(\Gamma_n \cap C_i \cap G_p)) \setminus (\varphi(\Gamma_n \cap G_{p+1/2^{i-1}}))\right) \qquad (19)$$

is empty if C_i does not contain segments "close" to $]x, y[$, and then the intersection among all the integers i is empty if C does not contain $]x, y[$; on the other hand, if C contains $]x, y[$ then the set in (19) is exactly $\varphi(]x, y[)$. The intersection for all the integers i, then, gives the thesis. We can now note that the same argument works in the general case with many maximal transport rays, since we can consider separately each maximal ray and apply the argument to it; note also that the role of the no-crossing condition (18) is only to ensure that each point in T_Γ is contained in a unique maximal transport ray, which allows to define the map π_r.

Now we state an "abstract" existence result on optimal transport maps. The result is valid under some regularity conditions on the decomposition in transport rays induced by Γ. Notice that assumption (ii) below on the μ-negligibility of left extreme points which are not fixed points is necessary for $n \geq 3$ even for absolutely continuous measures μ, in view of the counterexample in [33]. We will show in [5] that (ii), (iii) hold for $n = 2$ whenever μ is absolutely continuous.

Theorem 6.1. *Let $\mu, \nu \in \mathcal{P}(\mathbb{R}^n)$ with finite first order moments and let $\gamma \in \Pi(\mu, \nu)$ be concentrated on Γ. Assume that (18) holds and that*

(i) μ is absolutely continuous with respect to $\mathcal{L}^n$;
(ii) $T_\Gamma^l \setminus (T_\Gamma \cup F_\Gamma)$ is μ-negligible;
(iii) there exists a μ-negligible set $N \subset T_\Gamma$ and an increasing sequence of compact sets K_h such that $\tau_r|_{K_h}$ is a Lipschitz map and the union of K_h is $T_\Gamma \setminus N$.

Then there exists $\gamma_\# \in \Pi(\mu, \nu)$ such that

(a) $\gamma_\#$ is induced by a transport map t and t is optimal whenever Γ is c-monotone.
(b) $\gamma_\#$ is concentrated on the set of pairs (x, y) such that either $x = y$ or $[x, y]$ is contained in the closure of a maximal transport ray of Γ.
(c) For any convex function $\phi : [0, +\infty) \to \mathbb{R}$ bounded from below we have

$$\int_{\mathbb{R}^n \times \mathbb{R}^n} \phi(|x - y|) \, d\gamma_\# \leq \int_{\mathbb{R}^n \times \mathbb{R}^n} \phi(|x - y|) \, d\gamma.$$

If ϕ is strictly convex the inequality above is strict unless $\gamma = \gamma_\#$.

Proof. We split the proof in five steps. In the first four steps we assume that $F_\Gamma \subset T_\Gamma$, i.e. that any fixed point is contained in a transport ray. In the following, points in the closure of the same maximal transport rays will be ordered according to the orientation of the ray.

Step 1. The map r which associates to a pair $(x, y) \in \Gamma$ the closure of the maximal transport ray containing $]x, y[$ when $x \neq y$ and containing $\{x\}$ when $x = y$ is well defined on Γ, hence defined γ-a.e. According to Theorem 9.2 we can represent

$$\gamma = \gamma_C \otimes \sigma \qquad \text{with} \qquad \sigma := r_\# \gamma \in \mathcal{P}(\mathcal{S}_c(\mathbb{R}^n))$$

where γ_C are probability measures in $\mathbb{R}^n \times \mathbb{R}^n$ concentrated on $r^{-1}(C) \subset C \times C$, and therefore satisfying

$$\gamma_C \left(\{ (x, y) \in C \times C : y < x \} \right) = 0. \tag{20}$$

We define $\mu_C := \pi_{0\#} \gamma_C$ and $\nu_C := \pi_{1\#} \gamma_C$, so that $\gamma_C \in \Pi(\mu_C, \nu_C)$ and (37) yields

$$\mu = \mu_C \otimes \sigma, \qquad \nu = \nu_C \otimes \sigma. \tag{21}$$

Since $r^{-1}(C) \subset C \times C$ the probability measures μ_C, ν_C are concentrated on C. More precisely, we will use in the following the fact that μ_C is concentrated on $\pi_0 \left(r^{-1}(C) \right)$. Notice also that μ_C, ν_C have finite first order moments for σ-a.e. C because μ and ν have finite first order moments.

Step 2. We claim that μ_C has no atom for σ-a.e. C.

Taking into account Proposition 6.1(ii), the inclusion $F_\Gamma \subset T_\Gamma$ and the assumption (ii) we obtain that μ is concentrated on T_Γ, therefore μ_C is concentrated on $T_\Gamma \cap \pi_0(r^{-1}(C))$ for σ-a.e. C. Now we check that, due to condition (18), $T_\Gamma \cap \pi_0(r^{-1}(C))$ is contained in the relative interior of C. Indeed, let $x \in T_\Gamma \cap \pi_0(r^{-1}(C))$, let $(x', y') \in \Gamma$ such that $x \in]x', y'[$ and let C_0 be the closure of the maximal transport ray containing $]x', y'[$. If $C = C_0$ then x is in the relative interior of C; if, on the other hand, $C \neq C_0$ then there exists $y \in \mathbb{R}^n$ such that $r(x, y) = C$, thus condition (18) is violated because $[x, y]$ and $[x', y']$ intersect at x', in the relative interior of $[x, y]$.

Denoting by $\pi_r : T_\Gamma \mapsto S_o(\mathbb{R}^n)$ the map which associates to a point the maximal transport ray containing it, by Theorem 9.4, Remark 9.1 and assumption (iii) we have $\mu = \mu'_A \otimes \theta$ with $\theta = \pi_{r\#}\mu$, μ'_A concentrated on $\pi_r^{-1}(A) \subset A$, and $\mu'_A \ll \mathcal{H}^1 \llcorner A$ (and in particolar has no atom) for θ-a.e. A.

Denoting by $\mathrm{cl} : S_o(\mathbb{R}^n) \to S_c(\mathbb{R}^n)$ the bijection which associates to an open segment its closure, we have

$$\mu = \mu_C \otimes \sigma(C) = \mu_{\mathrm{cl}(A)} \otimes \mathrm{cl}_\#^{-1}\sigma(A)$$

and since $\mu_{\mathrm{cl}(A)}$ are probability measures concentrated on A the uniqueness Theorem 9.2 gives $\mathrm{cl}_\#^{-1}\sigma = \theta$ and $\mu_{\mathrm{cl}(A)} = \mu'_A$ for θ-a.e. A.

As $\sigma = \mathrm{cl}_\#\theta$, this proves that μ_C has no atom for σ-a.e. C.

Step 3. According to Theorem 5.1 we can find a non-decreasing map t_C defined on the relative interior of C (i.e. the maximal transport ray relative to C) and with values in C, such that $t_{C\#}\mu_C = \nu_C$. Moreover, by Theorem 5.2 and (20), for any convex function $\phi : [0, +\infty) \to \mathbb{R}$ bounded from below we have

$$\int_C \phi(|x - t_C(x)|)\, d\mu_C \leq \int_{C \times C} \phi(|x - y|)\, d\gamma_C \tag{22}$$

with strict inequality if ϕ is strictly convex and $(Id \times t_C)_\#\mu_C \neq \gamma_C$.

Step 4. We define t on T_Γ by gluing the maps t_C. Since the map $C \mapsto \gamma_C = (Id \times t_C)_\#\mu_C$ is Borel (as a measure-valued map, see the Appendix) by Theorem 9.3 we infer the existence of a Borel map t such that $t = t_C$ μ_C-a.e. for σ-a.e. C. As t_C and t map μ_C in ν_C, it follows immediately from (21) that t maps μ into ν.

Setting $\gamma_\# := (Id \times t)_\#\mu \in \Pi(\mu, \nu)$, conditions (a) and (b) are satisfied by construction, as the segments $[x, t(x)]$ are contained in the closure of a maximal transport ray of Γ. Condition (c) follows by (22) after an integration on $S_c(\mathbb{R}^n)$ with respect to σ.

Step 5. Now we remove the assumption that $F_\Gamma \subset T_\Gamma$. Let $L := F_\Gamma \setminus T_\Gamma$ and let $\Gamma' = \Gamma \setminus \{(x,x) : x \in L\}$. By applying the first four steps to $\gamma' := \gamma \llcorner \Gamma'$ we obtain a transport map t defined on $T_{\Gamma'}$ mapping $\pi_{0\#}\gamma'$ to $\pi_{1\#}\gamma'$. Noticing that $T_\Gamma = T_{\Gamma'}$, it suffices to extend t to L setting $t = Id$ on L and to set $\gamma_\# := (Id \times t)_\# \mu$.

Now we assume that $\|\cdot\|$ is a norm in $\mathbb{R}^n$ satisfying the regularity and uniform convexity conditions

$$c_1 \leq \frac{\partial^2}{\partial \xi \partial \xi} \|\cdot\|^2 \leq c_2 \quad \forall \xi \in \mathbf{S}^{n-1} \qquad \text{for some } c_2 \geq c_1 > 0. \qquad (23)$$

We consider a σ-compact set $\Gamma \subset \Gamma_u$, where

$$\Gamma_u := \{(x,y) \in \mathbb{R}^n \times \mathbb{R}^n : \|x - y\| = u(x) - u(y)\} \qquad (24)$$

for some function $u : \mathbb{R}^n \to \mathbb{R}$ which is 1-Lipschitz relative to the distance induced in $\mathbb{R}^n$ by $\|\cdot\|$.

Theorem 6.2. *With the choice of Γ above, (18) and the following properties hold:*

(i) For any maximal transport ray S we have $u(x') - u(y') = \|x' - y'\|$ whenever $x', y' \in S$ and $x' \leq y'$.
(ii) The set $T_\Gamma^l \setminus (T_\Gamma \cup F_\Gamma)$ is Lebesgue negligible.
(iii) Condition (iii) in Theorem 6.1 holds.

Proof. Arguing as in [16] (or [24] for the euclidean norm) one can show that at any point $z \in]x,y[$ inside T_Γ the function u is differentiable, $\|du(z)\|^* = 1$, and $(du(z))^*$ is the direction of the ray $]x,y[$. This immediately leads to the fact that (18) holds.
(i) We first show the property for transport rays. If $x', y' \in]x,y[$ and $x' \leq y'$ then

$$u(x) - u(x') \leq \|x - x'\| \qquad \text{and} \qquad u(y') - u(y) \leq \|y - y'\|,$$

so that, taking into account that $u(x) - u(y) = \|x - y\|$ and

$$\|x - y\| = \|x - x'\| + \|x' - y'\| + \|y' - y\|$$

we obtain that $u(x') - u(y') \geq \|x' - y'\|$. The extension to maximal transport rays is analogous.
Properties (ii), (iii) with $\Gamma = \Gamma_u$ and $N = \emptyset$ are shown in [16]. A fortiori (iii) holds when $\Gamma \subset \Gamma_u$, as $T_\Gamma \subset T_{\Gamma_u}$, while the set in (ii) is more sensitive to the choice of Γ. For the reader's convenience we give a different proof of both (ii) and (iii), in the spirit of [3] and based on semiconcavity estimates, in the general situation when Γ is contained in Γ_u. In the following we denote by T the set $T_{\Gamma_u}^l \setminus \Sigma$, where Σ is the $\mathcal{L}^n$-negligible Borel set where u is not differentiable.

Step 1. We show that $x \mapsto du(x)$ is $\mathcal{L}^n$-countably Lipschitz on T. Given a direction $\xi \in \mathbf{S}^{n-1}$ and $a \in \mathbb{R}$, let R be the union of the half closed maximal transport rays $[x, y[$ with $\langle y - x, \xi \rangle \geq 0$ and $\langle y, \xi \rangle \geq a$. It suffices to prove that the restriction of du to

$$R_a := R \cap \{x : x \cdot \xi < a\}$$

has the countable Lipschitz property stated in the theorem. To this aim, since BV_{loc} funtions have this property (see for instance Theorem 5.34 of [2] or [26]), it suffices to prove that ∇u coincides $\mathcal{L}^n$-a.e. in R_a with a suitable function $w \in [BV_{\mathrm{loc}}(S_a)]^n$, where $S_a = \{x : x \cdot \xi < a\}$. To this aim we define

$$\tilde{u}(x) := \min \{u(y) + \|x - y\| : y \in Y_a\}$$

where Y_a is the collection of all right endpoints of maximal transport rays with $y \cdot \xi \geq a$. By construction $\tilde{u} \geq u$ and equality holds on R_a.

We claim that, for $b < a$, $\tilde{u} - C|x|^2$ is concave in S_b for $C = C(b)$ large enough. Indeed, since $\|x - y\| \geq a - b > 0$ for any $y \in Y_a$ and any $x \in S_b$, the functions

$$x \mapsto u(y) + \|x - y\| - C|x|^2, \qquad y \in Y_a$$

are all concave in S_b for C large enough depending on $a - b$ (here we use the upper estimate in (23)). In particular, as gradients of real valued concave functions are BV_{loc} (see for instance [1]), we obtain that

$$w := \nabla \tilde{u} = \nabla (\tilde{u} - C|x|^2) + 2Cx$$

is a BV_{loc} function in S_a. Since $\nabla u = w$ $\mathcal{L}^n$-a.e. in R_a the proof is achieved.

Now we notice that the duality map which associates to a unit vector $L \in (\mathbb{R}^n)^*$ the unique unit vector $v \in \mathbb{R}^n$ such that $L(v) = 1$ is Lipschitz (here we use the lower bound in (23)): indeed, setting $\phi(v) = \|v\|^2/2$, by the Lagrange multiplier rule we have $L + \lambda \nabla \phi(v) = 0$ for some $\lambda \in \mathbb{R}$ and evaluation at v gives $\lambda = -1/\langle \nabla \phi(v), v \rangle$ (because $\langle L, v \rangle = 1$), therefore

$$L = \frac{\nabla \phi(v)}{\langle \nabla \phi(v), v \rangle} = \frac{\nabla \phi(v)}{2\phi(v)} = \nabla \phi(v).$$

Since $v \mapsto \nabla \phi(v)$ is a strictly monotone operator its inverse is a Lipschitz map. It follows that $x \mapsto (du(x))^*$ is $\mathcal{L}^n$-countably Lipschitz on T.

Step 2. Now we show that the family L of left extreme points of maximal transport rays of Γ_u is Lebesgue negligible. As Γ_u is closed and

$$L = \bigcup_{(x,y) \in \Gamma_u} [x, y[\, \Big\backslash \bigcup_{(x,y) \in \Gamma_u}]x, y[$$

we have that L is a Borel set. For any $x \in L \setminus \Sigma$ there is a unique maximal transport ray emanating from x, with direction $(du(x))^*$ and with length

$l(x)$. It is immediate to check that l is upper semicontinuous on $L \setminus \Sigma$, and therefore l is a Borel function. By Lusin theorem it will be sufficient to show that $\mathcal{L}^n(K) = 0$ for any compact set $K \subset L \setminus \Sigma$ where l and $(du)^*$ are continuous. We define

$$B := \bigcup_{x \in K} [x, x + \frac{l(x)}{2}(du(x))^*] \setminus \Sigma$$

(B is Borel due to the continuity of l and of $(du)^*$) and we apply Theorem 9.4 with $\lambda = \mathcal{L}^n \llcorner K$ and $\tau(x) = (du(x))^*$ (notice that condition (iii) of the theorem holds because of Step 1 and the inclusion $B \subset T$) to get

$$\mathcal{L}^n(K) = \int_{\mathcal{S}_c(\mathbb{R}^n)} \lambda_C(K) \, d\mu(C) = 0$$

because $\lambda_C \ll \mathcal{H}^1 \llcorner C$ for μ-a.e. C and $K \cap C$ contains only one point for any closed maximal transport ray C.

Step 3. We show that $T^l_\Gamma \setminus (T_\Gamma \cup F_\Gamma)$ is Lebesgue negligible. Any point in this set is either a left extreme point of a maximal transport ray of Γ_u or is contained in T_{Γ_u}. Therefore, by Step 2, it suffices to consider only the set

$$R := T^l_\Gamma \cap T_{\Gamma_u} \setminus (T_\Gamma \cup F_\Gamma).$$

Since the intersection of R with any maximal transport ray of Γ_u is at most countable, by Remark 9.1 with $B = R$ and $\lambda = \mathcal{L}^n \llcorner R$ we obtain as in Step 2 that $\mathcal{L}^n(R) = 0$.

7 A stability result

In this section we assume that:

(I) μ, ν are probability measures in $\mathbb{R}^n$ with finite first order moments and $\mu \ll \mathcal{L}^n$;

(II) $c(x, y) = \|x - y\|$, where the norm $\|\cdot\|$ satisfies the regularity and uniform convexity conditions (23).

The main results of this section is the following existence and stability theorem. For $\varepsilon > 0$, we consider nondecreasing and strictly convex maps $\phi_\varepsilon : [0, +\infty) \to [0, +\infty)$ satisfying the following conditions:

(a) $\varepsilon \to \phi_\varepsilon(d)$ is convex and $\phi_\varepsilon(d) \to d$ as $\varepsilon \downarrow 0$ for any $d \geq 0$;

(b) the right derivative

$$\phi(d) := \lim_{\varepsilon \to 0^+} \frac{\phi_\varepsilon(d) - d}{\varepsilon}$$

exists and is a real-valued strictly convex function in $[0, +\infty)$ bounded from below.

Two model cases are $\phi_\varepsilon(d) = d + \varepsilon \phi(d)$, with ϕ strictly convex and bounded from below, or $\phi_\varepsilon(d) = d^{1+\varepsilon}$. In the latter case, $\phi(d) = d \ln d$.

Theorem 7.1. *Assume (I) and (II). Then:*

(i) The problem

$$\min \left\{ \int_{\mathbb{R}^n} \|t(x) - x\| \, d\mu(x) : t_\# \mu = \nu \right\} \qquad (25)$$

has a solution.

(ii) Assume that

$$\int_{\mathbb{R}^n \times \mathbb{R}^n} \phi_{\varepsilon_0}(\|x - y\|) \, d\gamma < \infty \qquad \text{for any } \gamma \in \Pi(\mu, \nu) \qquad (26)$$

for some $\varepsilon_0 > 0$ and let, for $\varepsilon \in (0, \varepsilon_0)$, t_ε be the optimal maps in the problem

$$\min \left\{ \int_{\mathbb{R}^n} \phi_\varepsilon (\|t(x) - x\|) \, d\mu(x) : t_\# \mu = \nu \right\}. \qquad (27)$$

Then $t_\varepsilon \to t$ in measure as $\varepsilon \to 0^+$ and t solves (25).

Proof. We need only to prove statement (ii). Indeed, choosing $\phi(t) + \sqrt{1 + t^2} - 1$, (26) holds for any $\varepsilon_0 > 0$ with $\phi_\varepsilon = Id + \varepsilon\phi$. Therefore we can apply statement (ii) to obtain an optimal transport map as limit as $\varepsilon \to 0^+$ of maps t_ε.

The proof of (ii) relies essentially on Theorem 7.2 below. Indeed, the variational argument of Proposition 7.1, based on the theory of Γ-asymptotic expansions, shows that any γ_0, limit point of $\gamma_\varepsilon = (Id \times t_\varepsilon)_\# \mu$ as $\varepsilon \to 0^+$, is an optimal planning for the Kantorovich problem and, in addition, minimizes the secondary variational problem

$$\gamma \mapsto \int_{\mathbb{R}^n \times \mathbb{R}^n} \phi(\|x - y\|) \, d\gamma$$

among all optimal plannings for the primary variational problem. Theorem 7.2 says that there is a unique such minimizer induced by a transport map t, i.e. $\gamma_0 = (Id \times t)_\# \mu$.

This shows that problem (25) has a solution and that $(Id \times t_\varepsilon)_\# \mu \to (Id \times t)_\# \mu$ weakly as $\varepsilon \to 0^+$. Let now $\delta > 0$ and choose a compact set $K \subset \mathbb{R}^n$ such that $t|_K$ is continuous and $\mu(\mathbb{R}^n \setminus K) < \delta$. Denoting by $\tilde{t}$ a continuous extension of $t|_K$ and choosing as test function

$$\varphi(x, y) = \chi_K(x) \times \psi (y - \tilde{t}(x))$$

with $\psi \in C(\mathbb{R}^n, [0, 1])$, $\psi(0) = 0$ and $\psi(z) = 1$ for $|z| \geq \delta$, we obtain

$$\limsup_{\varepsilon \to 0^+} \mu (\{|t_\varepsilon - t| > \delta\}) \leq \delta + \limsup_{\varepsilon \to 0^+} \mu (K \cap \{|t_\varepsilon - t| > \delta\})$$

$$\leq \delta + \limsup_{\varepsilon \to 0^+} \int \varphi \, d\gamma_\varepsilon$$

$$\leq \delta + \int \varphi \, d(Id \times t)_\# \mu = \delta.$$

Since $\delta > 0$ is arbitrary this proves the convergence in measure of t_ε to t.

Theorem 7.2. *Assume (I) and (II). Let $\Pi_1(\mu, \nu)$ be the collection of all optimal plannings in the primary problem*

$$\min \left\{ \int_{\mathbb{R}^n \times \mathbb{R}^n} \|x - y\| \, d\gamma : \ \gamma \in \Pi(\mu, \nu) \right\} \tag{28}$$

and let $\phi : [0, +\infty) \to \mathbb{R}$ be a strictly convex function bounded from below. Let us consider the secondary variational problem

$$\min \left\{ \int_{\mathbb{R}^n \times \mathbb{R}^n} \phi(\|x - y\|) \, d\gamma : \ \gamma \in \Pi_1(\mu, \nu) \right\} \tag{29}$$

and let us assume that the minimum is finite. Then (29) has a unique solution and this solution is induced by a transport map t.

Proof. The existence of a solution γ_0 of the secondary variational problem is a direct consequence of the weak compactness of the class $\Pi_1(\mu, \nu)$. As c-concavity reduces to 1-Lipschitz continuity when the cost function is a distance, according to Theorem 3.3 there exists a function $u : \mathbb{R}^n \to \mathbb{R}$ which is 1-Lipschitz with respect to the distance induced by $\| \cdot \|$ and such that $\gamma \in \Pi_1(\mu, \nu)$ if and only if

$$\mathrm{spt}\, \gamma \subset \{ (x, y) \in \mathbb{R}^n \times \mathbb{R}^n : \ u(x) - u(y) = \|x - y\| \}.$$

We define $\Gamma = \Gamma_u$ as in (24) and we wish to apply Theorem 6.1 to γ_0. Obviously assumption (i) of the theorem is satisfied, while assumptions (ii), (iii) follow by Theorem 6.2(ii) and Theorem 6.2(iii).

By Theorem 6.1 we obtain $\gamma_\# = (Id \times t)_\# \mu \in \Pi(\mu, \nu)$ such that

$$\int_{\mathbb{R}^n \times \mathbb{R}^n} \psi(|x - y|) \, d\gamma_\# \leq \int_{\mathbb{R}^n \times \mathbb{R}^n} \psi(|x - y|) \, d\gamma$$

for any convex function $\psi : [0, +\infty) \to \mathbb{R}$ bounded from below, with strict inequality if ψ is strictly convex. Choosing $\psi(t) = t$ we obtain that $\gamma_\# \in \Pi_1(\mu, \nu)$. Choosing $\psi = \phi$ the minimality of γ_0 in (29) gives $\gamma_0 = \gamma_\#$, and therefore γ_0 is induced by a transport map.

The uniqueness of γ_0 is an easy consequence of the linear structure of the variational problems (28), (29): if $\gamma_0 = (Id \times t)_\# \mu$ and $\gamma_0' = (Id \times t')_\# \mu$ are both optimal then $(\gamma_0 + \gamma_0')/2$ is still optimal and therefore is induced by a transport map. This is possible only if $t = t'$ μ-a.e.

Remark 7.1. The secondary variational problem (29) can also be rephrased as follows: minimize

$$\gamma \mapsto \int_{\mathbb{R}^n \times \mathbb{R}^n} \tilde{c}(x, y) \, d\gamma$$

in $\Pi(\mu, \nu)$, where $\tilde{c} : \mathbb{R}^n \times \mathbb{R}^n \to [0, +\infty]$ is given by

$$\tilde{c}(x,y) = \begin{cases} \phi(\|x - y\|) & \text{if } \|x - y\| \le u(x) - u(y) \\ +\infty & \text{if } \|x - y\| > u(x) - u(y). \end{cases}$$

Indeed, the duality theory for the primary variational problem says that $\gamma \in \Pi_1(\mu, \nu)$ if and only if spt $\gamma \subset \Gamma_u$.

Proposition 7.1. *Assume (I) and (26). For $\varepsilon \in (0, \varepsilon_0)$, let t_ε be the optimal maps in (27) and let $\gamma_\varepsilon = (Id \times t_\varepsilon)_{\#}\mu$ be the optimal plannings associated to t_ε. Then any limit point γ_0 of γ_ε is a minimizer of the secondary variational problem (29) and the minimum if finite.*

Proof. For $\gamma \in \Pi(\mu, \nu)$ and $\varepsilon \in (0, \varepsilon_0)$ we define

$$F_\varepsilon(\gamma) := \int_{\mathbb{R}^n \times \mathbb{R}^n} \phi_\varepsilon\left(\|x - y\|\right) d\gamma, \qquad F(\gamma) := \int_{\mathbb{R}^n \times \mathbb{R}^n} \|x - y\| \, d\gamma.$$

Let $m = \min F$ and $F'_\varepsilon := (F_\varepsilon - m)/\varepsilon$. According to Theorem 4.1 it suffices to show that F_ε Γ-converge in $\Pi(\mu, \nu)$ to F and F'_ε Γ-converge in $\Pi(\mu, \nu)$ to

$$F'(\gamma) := \begin{cases} \displaystyle\int_{\mathbb{R}^n \times \mathbb{R}^n} \phi\left(\|x - y\|\right) d\gamma & \text{if } \gamma \in \Pi_1(\mu, \nu) \\ +\infty & \text{otherwise} \end{cases}$$

(here we consider any metric in $\Pi(\mu, \nu)$ inducing the convergence in the duality with $C_b(\mathbb{R}^n \times \mathbb{R}^n)$).

In order to show the Γ-convergence of F_ε to F, we notice that the convexity of $\varepsilon \mapsto \phi_\varepsilon(d)$ gives

$$\liminf_{\varepsilon \to 0^+} F_\varepsilon(\gamma_\varepsilon) \ge \liminf_{\varepsilon \to 0^+} \int_B \|y - x\| - \varepsilon\phi\left(\|y - x\|\right) d\gamma_\varepsilon \ge \int_B \|y - x\| \, d\gamma$$

for any family γ_ε weakly converging to γ and any bounded open set $B \subset \mathbb{R}^n \times \mathbb{R}^n$. Letting $B \uparrow \mathbb{R}^n \times \mathbb{R}^n$ the lim inf inequality follows.

The lim sup inequality with $\gamma_\varepsilon = \gamma$ is again a direct consequence of a convexity argument, which provides the estimate

$$F_\varepsilon(\gamma) - F(\gamma) \le \frac{\varepsilon}{\varepsilon_0} \left(F_{\varepsilon_0}(\gamma) - F(\gamma)\right).$$

Now we show the Γ-convergence of F'_ε, starting from the lim inf inequality. Let $\gamma_\varepsilon \to \gamma$ weakly and assume with no loss of generality that $\liminf_\varepsilon F'_\varepsilon(\gamma_\varepsilon)$ is finite. Then, the Γ-convergence of F_ε to F gives that $\gamma \in \Pi_1(\mu, \nu)$ and the convexity of the map $\varepsilon \mapsto \phi_\varepsilon(d)$ gives $\phi_\varepsilon(d) \ge d + \varepsilon\phi(d)$, so that

$$F_\varepsilon(\gamma_\varepsilon) - m \ge F_\varepsilon(\gamma_\varepsilon) - F(\gamma_\varepsilon) \ge \varepsilon F'(\gamma_\varepsilon).$$

As ϕ is continuous and bounded from below, dividing both sides by ε we obtain

$$\liminf_{\varepsilon \to 0^+} F'_\varepsilon(\gamma_\varepsilon) \geq \liminf_{\varepsilon \to 0^+} F'(\gamma_\varepsilon) \geq F'(\gamma).$$

In order to show the lim sup inequality we can assume with no loss of generality that $\gamma \in \Pi_1(\mu, \nu)$. Then

$$\frac{F_\varepsilon(\gamma) - m}{\varepsilon} = \int_{\mathbb{R}^n \times \mathbb{R}^n} \frac{\phi_\varepsilon(\|y - x\|) - \|y - x\|}{\varepsilon} \, d\gamma$$

and since $\varepsilon \mapsto (\phi_\varepsilon(d) - d)/\varepsilon$ is nondecreasing the dominated convergence theorem gives

$$\lim_{\varepsilon \to 0^+} \frac{F_\varepsilon(\gamma) - m}{\varepsilon} = F'(\gamma).$$

8 A counterexample

In the transport problem in the Euclidean space $\mathbb{R}^n$, the condition

$$\mu(B) = 0 \quad \text{whenever} \quad B \in \mathcal{B}(\mathbb{R}^n) \text{ and } \mathcal{H}^{n-1}(B) < \infty \tag{30}$$

ensures existence of optimal transport maps whenever ν has compact support and $c(x, y) = h(y - x)$, with h *strictly* convex and locally $C^{1,1}$ (see [30]). Condition (30) is sharp, as the following simple and well-known example shows:

Example 8.1. Let $I = [0, 1]$, $\mu = \mathcal{H}^1 \llcorner \{0\} \times I$, $2\nu = \mathcal{H}^1 \llcorner \{-1\} \times I + \mathcal{H}^1 \llcorner \{1\} \times I$. Then, choosing $c(x, y) = |x - y|^\alpha$, with $\alpha > 0$, it is easy to check that $\gamma_x := (\delta_{(1, x_2)} + \delta_{(-1, x_2)})/2$ is the unique solution of the Kantorovich problem. Therefore the classical transport problem has no solution.

In this section we show that (30) does not provide in general existence of optimal transport maps when the cost function is the euclidean distance, building measures μ with dimension arbitrarily close to n such that the transport problem has no solution. This basically happens because in this case different maximal transport rays cannot cross in their interior (see (18)). Our construction provides also a counterexample to the statement made in the last page of [41] about the existence of optimal transport maps for measures μ such that $\mu(B_r(x)) = o(r^{n-1})$.

Lemma 8.1 (Horizontal transport rays). *Let* $I = [0, 1]$ *and let* μ, ν *be probability measures in* $\mathbb{R}^2$ *with support respectively in* $I \times I$ *and* $[5, 6] \times I$. *We assume that*

$$\mu([0, 1] \times [0, t]) = \nu([5, 6] \times [0, t]) \qquad \forall t \in I. \tag{31}$$

Then the optimal plannings move mass only along horizontal rays.

Proof. Let γ be an optimal planning relative to μ, ν. Assuming by contradiction that mass is not transported along horizontal rays, the number

$$\varepsilon := \sup\{|y_2 - x_2| : (x, y) \in \operatorname{spt}\gamma\}$$

is strictly positive and we can choose $(x, y) \in \operatorname{spt}\gamma$ such that $|y_2 - x_2| > \varepsilon/2$. We assume with no loss of generality (up to a reflection) that $y_2 < x_2$ and we prove by an elementary geometric argument the existence of $(x', y') \in \operatorname{spt}\gamma$ with $y_2' - x_2' > \varepsilon$, thus reaching a contradiction.

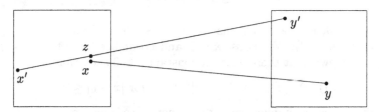

Fig. 2. Location of rays in Lemma 8.1

Applying the mass balance condition (31) with $t = (x_2 + y_2)/2$, we can find $(x', y') \in \operatorname{spt}\gamma$ such that $y_2' \geq t \geq x_2'$ and, since the rays $[x, y]$ and $[x', y']$ cannot cross, there is $z \in [x', y']$ with $z_1 = x_1$ and $z_2 > x_2$ as in Figure 2 (possibly exchanging the roles of μ and ν). By easy geometric arguments the following inequalities hold:

$$z_1 - x_1' \leq 1, \qquad y_1' - x_1' \geq 4, \qquad z_2 - x_2' \geq x_2 - t = \frac{x_2 - y_2}{2} > \frac{\varepsilon}{4}.$$

Then, since by similitude $\dfrac{y_2' - x_2'}{z_2 - x_2'} = \dfrac{y_1' - x_1'}{z_1 - x_1'}$, it follows $y_2' - x_2' > \varepsilon$, the searched contradiction.

Theorem 8.1. *For any $s \in (1, 2)$ there exists a continuous function $f : I \to I$ such that, setting*

$$\Gamma := \{(x_1, x_2) : x_1 = f(x_2)\}, \quad \mu := (f \times Id)_{\#}\mathcal{L}^1 \llcorner I, \quad \nu := \mathcal{L}^2\llcorner([5, 6] \times I)$$

the following properties hold:

(i) The Hausdorff dimension of Γ is s and $\mu \ll \mathcal{H}^t$ for any $t < s$;
(ii) the Kantorovich problem with data μ, ν and $c(x, y) = |x - y|$ has the unique solution $\gamma_x = \mathcal{H}^1\llcorner([5, 6] \times \{x_2\})$.

In particular μ satisfies (30) but the classical optimal transport problem has no solution.

Proof. (i) We use the classical construction given in Theorem 8.2 of [25]. Let $g : \mathbb{R} \to \mathbb{R}$ be the 4-periodic sawtooth function defined by

$$g(x) = \begin{cases} x & \text{if } 0 \leq x \leq 1 \\ 2 - x & \text{if } 1 \leq x \leq 3 \\ x - 4 & \text{if } 3 \leq x \leq 4 \end{cases} \tag{32}$$

and set

$$f(x) := \kappa + \sum_{i=1}^{\infty} \lambda_i^{s-2} g(\lambda_i x) \qquad x \in I$$

where $(\lambda_i) \subset (0, +\infty)$ are such that $\lambda_{i+1}/\lambda_i \to +\infty$ and $\ln(\lambda_{i+1})/\ln(\lambda_i) \to 1$ (for instance $\lambda_i = i!$). We choose $\kappa \in \mathbb{R}$ and normalize λ_i so that $0 \leq f \leq 1$. In [25] it is shown that there exists a constant $\delta > 0$ such that

$$|f(x) - f(y)| \leq 6|x - y|^{2-s} \qquad \text{for } |x - y| \leq \delta. \tag{33}$$

As a consequence, a simple covering argument (see Theorem 8.1 in [25]) gives

$$\mathcal{H}^s \left(\Gamma \cap Q_r(x) \right) \leq c r^s \qquad \forall r \in (0, 2), \tag{34}$$

with $c = c(s, \delta)$, for any cube $Q_r(x)$ with side length r centered at $x \in \Gamma$. In particular $\mathcal{H}^s(\Gamma) < \infty$.

Another estimate still proved in [25] (see (8.12) on page 117) gives for any $t < s$ the existence of a constant $c_1 = c_1(t)$ such that

$$\mu \left(\Gamma \cap Q_r(x) \right) \leq c_1 r^t \qquad \forall x \in \Gamma, \ r \in (0, 2). \tag{35}$$

It follows that $\mu \ll \mathcal{H}^t$ for any $t < s$. If $\mathcal{H}^t(\Gamma)$ were finite for some $t < s$ then $\mathcal{H}^{t'}(\Gamma)$ would be equal to 0 for $t' = (s+t)/2$, hence $\mu(\Gamma)$ would be zero. This contradiction proves that $\mathcal{H}^t(\Gamma) = +\infty$ for any $t < s$, hence the Hausdorff dimension of Γ is s.

(ii) The measures μ, ν satisfy by construction the identity (31). By Lemma 8.1 the support of γ_x is contained in $[5, 6] \times \{x_2\}$ for μ-a.e. x. Since

$$\nu = \int_{\mathbb{R}^2} \gamma_x \, d\mu(x) = \int_I \gamma_{(f(t), t)} \, dt$$

and the measures $\gamma_{(f(t), t)}$ are supported on $\mathbb{R} \times \{t\}$, the uniqueness of the disintegration of ν with respect to $t = x_2$ yields $\gamma_{(f(t), t)} = \mathcal{H}^1 \llcorner ([5, 6] \times \{t\})$ for a.e. $t \in I$.

9 Appendix: disintegration of measures

In this appendix we recall some basic facts about disintegration of measures, focussing for simplicity on the case of positive measures. In this section, unless otherwise stated, all spaces X, Y, Z we consider are locally compact and separable metric spaces.

Theorem 9.1 (Existence). *Let $\pi : X \to Y$ be a Borel map, let $\lambda \in \mathcal{M}_+(X)$ and set $\mu = \pi_\# \lambda \in \mathcal{M}_+(Y)$. Then there exist measures $\lambda_y \in \mathcal{M}_+(X)$ such that*

(i) $y \mapsto \lambda_y$ is a Borel map and $\lambda_y \in \mathcal{P}(X)$ for μ-a.e. $y \in Y$;
(ii)$\lambda = \lambda_y \otimes \mu$, i.e.

$$\lambda(A) = \int_Y \lambda_y(A)\, d\mu(y) \qquad \forall A \in \mathcal{B}(X); \qquad (36)$$

(iii)λ_y is concentrated on $\pi^{-1}(y)$ for μ-a.e. $y \in Y$.

According to our terminology (which maybe is not canonical), a map $y \mapsto \lambda_y$ is Borel if $y \mapsto \lambda_y(B)$ is a Borel map in Y for any $B \in \mathcal{B}(X)$. This is equivalent (see for instance [2]) to the property that

$$y \mapsto \int_X \varphi(x, y)\, d\lambda_y(x)$$

is a Borel map in Y for any bounded Borel function $\varphi : X \times Y \to \mathbb{R}$. Our terminology is also justified by the observation that $y \mapsto \lambda_y$ is a Borel map in the conventional sense if we view $\mathcal{P}(X)$ as a subset of the compact metric of all positive Radon measures with total mass less than 1, endowed with the weak* topology coming from the duality with continuous and compactly supported functions in X. We will always make this embedding when we need to consider the space of probability measures as a measurable space.

The representation provided by Theorem 9.1 of λ can be used sometimes to compute the push forward of λ. Indeed,

$$f_\#(\lambda_y \otimes \mu) = f_\# \lambda_y \otimes \mu \qquad (37)$$

for any Borel map $f : X \to Z$, where Z is any other metric space. Notice also that if $T : Y \to Z$ is a Borel and 1-1 map, $\mu' := T_\# \mu$ and $\lambda = \eta_z \otimes \mu'$, then

$$\lambda_y = \eta_{T(y)} \quad \text{for } \mu\text{-a.e. } y \in Y. \qquad (38)$$

This is a simple consequence of the uniqueness of the disintegration, see Theorem 9.2 below.

The proof of Theorem 9.1 is available in many textbooks of measure theory or probability (in this case λ_y are the the so-called conditional probabilities induced by the random variable π, see for instance [2, 19]). In the case when $X = Y \times Z$ is a product space and $\pi(y, z) = y$ is the projection on the first variable the measures λ_y are concentrated on $\pi^{-1}(y) = \{y\} \times Z$, therefore it is often convenient to consider them as measures on Z, rather than measures on X, writing (36) in the form

$$\lambda(B) = \int_Y \lambda_y(\{z : (y, z) \in B\})\, d\mu(y) \qquad \forall B \in \mathcal{B}(X). \qquad (39)$$

We will always use this convention in the Kantorovich problem, writing each $\gamma \in \Pi(\mu, \nu) \subset \mathcal{P}(X \times Y)$ as $\gamma_x \otimes \mu$ with γ_x probability measures in Y.

Once the decomposition theorem is known in the special case $X = Y \times Z$ and $\pi(y, z) = z$ the general case can be easily recovered: it suffices to embed X into the product $Y \times X$ through the map $f(x) = (\pi(x), x)$ and to apply the decomposition theorem to $\tilde{\lambda} = f_\# \lambda$.

Now we discuss the uniqueness of λ_y and μ in the representation $\lambda = \lambda_y \otimes \mu$.

Theorem 9.2 (Uniqueness). *Let X, Y and π be as in Theorem 9.1; let $\lambda \in \mathcal{M}_+(X)$, $\mu \in \mathcal{M}_+(Y)$ and let $y \mapsto \eta_y$ be a Borel $\mathcal{M}_+(X)$-valued map defined on Y such that*

(i) $\lambda = \eta_y \otimes \mu$, i.e. $\lambda(A) = \int_Y \eta_y(A) \, d\mu(y)$ for any $A \in \mathcal{B}(X)$;
(ii) η_y is concentrated on $\pi^{-1}(y)$ for μ-a.e. $y \in Y$.

Then the η_y are uniquely determined μ-a.e. in Y by (i), (ii) and moreover, setting $C = \{y : \eta_y(X) > 0\}$, the measure $\mu \llcorner C$ is absolutely continuous with respect to $\pi_\# \lambda$. In particular

$$\frac{\mu \llcorner B}{\pi_\# \lambda} \eta_y = \lambda_y \qquad \text{for } \pi_\# \lambda\text{-a.e. } y \in Y \tag{40}$$

where λ_y are as in Theorem 9.1.

Proof. Let η_y, η_y' be satisfying (i), (ii). We have to show that $\eta_y = \eta_y'$ for μ-a.e. y. Let (A_n) be a sequence of open sets stable by finite intersection which generates the Borel σ-algebra of X. Choosing $A = A_n \cap \pi^{-1}(B)$, with $B \in \mathcal{B}(Y)$, in (i) gives

$$\int_B \eta_y(A_n) \, d\mu(y) = \int_B \eta_y'(A_n) \, d\mu(y).$$

Being B arbitrary, we infer that $\eta_y(A_n) = \eta_y'(A_n)$ for μ-a.e. y, and therefore there exists a μ-negligible set N such that $\eta_y(A_n) = \eta_y'(A_n)$ for any $n \in \mathbf{N}$ and any $y \in Y \setminus N$. By a well-know coincidence criterion for measures (see for instance Proposition 1.8 of [2]) we obtain that $\eta_y = \eta_y'$ for any $y \in Y \setminus N$.

Let $B' \subset B$ be any $\pi_\# \lambda$-negligible set; then $\pi^{-1}(B')$ is λ-negligible and therefore (ii) gives

$$0 = \int_Y \eta_y \left(\pi^{-1}(B') \right) \, d\mu(y) = \int_{B'} \eta_y(X) \, d\mu(y).$$

As $\eta_y(X) > 0$ on $B \supset B'$ this implies that $\mu(B') = 0$. Writing $\mu \llcorner B = h \pi_\# \lambda$ we obtain $\lambda = h \eta_y \otimes \pi_\# \lambda$ and $\lambda = \lambda_y \otimes \pi_\# \lambda$. As a consequence (40) holds.

In the following proposition we address the delicate problem, which occurs in optimal transport problems, of the measurability of maps f obtained by "gluing" different transport maps defined on the level sets of π. Simple examples show that the λ_y-measurability of f, though necessary, it is not sufficient:

for instance when $X = Y \times Z$ is a product space, π is the projection on the first factor and λ is concentrated on the graph of $\phi : Y \to Z$, then λ_y are Dirac masses concentrated at $(y, \phi(y))$ for μ-a.e. y, therefore λ_y-measurabilty provides no information on λ-measurability or, rather, on the existence of a Borel map g such that $g = f$ λ_y-a.e. for μ-a.e. $y \in Y$.

In order to state our measurability criterion we need some more terminology. Given a λ_y-measurable function $f : X \to Z$, we canonically associate to f the measure $\gamma_y(f) := (Id \times f)_\# \lambda_y$, a probability measure in $X \times Z$. It turns out that the measurability of the map $y \mapsto \gamma_y$ is sufficient to provide a Borel map g equivalent to f, i.e. such that $g = f$ λ_y-a.e. for μ-a.e. y.

Theorem 9.3 (Measurability criterion). *Keeping the notation of Theorem 9.1, let $f : X \to Z$ be satisfying the following two conditions:*

(i) f is λ_y-measurable for μ-a.e. y;
(ii) $y \mapsto \gamma_y(f)$ is a μ-measurable map between Y and $\mathcal{P}(X \times Z)$.

Then there exists a Borel map $g : X \to Z$ such that $g = f$ λ_y-a.e. for μ-a.e. y.

Proof. For the sake of simplicity we consider only the case when X, Y, Z are compact. Given $\nu \in \mathcal{M}_+(X)$ we define a metric on the space $L(X, \nu; Z)$ of ν-measurable maps between X and Z by

$$d_\nu(f, g) := \int_X d_Z\left(f(x), g(x)\right) \, d\nu(x).$$

It is well known that this metric induces the convergence in ν-measure and that $L(X, \nu; Z)$ is a complete metric space (with the canonical equivalence relation between ν-measurable maps).

Without loss of generality (by Lusin's theorem, see Theorem 2.3.5 in [26]) we can assume that:
(a) π is continuous (so that spt $\lambda_y \subset \pi^{-1}(y)$ are pairwise disjoint).
(b) $y \mapsto$ spt λ_y is continuous between Y and the class $\mathcal{K}$ of closed subsets of X, endowed with the Hausdorff metric. Indeed, the σ-algebra of the Borel subsets of $\mathcal{K}$ is generated by the sets $\{K : K \cap U \neq \emptyset\}$, for $U \subset X$ open, and

$$\{y : \text{spt } \lambda_y \cap U \neq \emptyset\} = \{y : \lambda_y(U) > 0\} \in \mathcal{B}(Y).$$

(c) $y \mapsto \lambda_y$ is continuous.
Step 1. We assume first that the restriction of f to spt λ_y is a M-Lipschitz function with values in Z for some $M \geq 0$ independent of y. By Lusin's theorem again we can find an increasing family of compact sets $Y_h \subset Y$ whose union covers μ-almost all of Y and such that $\gamma_y(f)$ restricted to Y_h is continuous for any h.

We claim now that the restriction of f to the compact set (compactness comes from assumption (b))

$$X_h := \bigcup_{y \in Y_h} \text{spt } \lambda_y$$

is continuous. Indeed, assume that $x_k \in \text{spt } \lambda_{y_k}$ converge to $x \in \text{spt } \lambda_y$ (with $y_k, y \in Y_h$) but $d_Y(f(x_k), f(x)) \geq \epsilon$ for some $\epsilon > 0$. Then the equi-Lipschitz condition provides $r > 0$ such that $d_Y(f(z), f(w)) \geq \epsilon/2$ for any choice of $z \in B_r(x) \cap \text{spt } \lambda_{y_k}$, $w \in B_r(x) \cap \text{spt } \lambda_y$. Choosing a test function χ of the form $\chi_1(x)\chi_2(z)$, with $\text{spt } \chi_1 \subset B_r(x)$, $\text{spt } \chi_2 \subset B_{\epsilon/2}(f(x))$ and $\chi_2(f(x)) = 1$ we find

$$\lim_{k \to +\infty} \gamma_{y_k}(\chi) = 0 < \gamma_y(\chi),$$

contradicting the continuity of $y \mapsto \gamma_y$ on Y_h.

As the union of X_h covers λ-almost all of X we obtain that f is λ-measurable.

Step 2. Now we attack the general case. For any h and any $y \in Y$ we consider the set $K_h(y)$ of all probability measures in $X \times Z$ of the form $(Id \times f)_{\#}\lambda_y$, with $f : X \to Z$ with Lipschitz constant less than h. By assumption (c) the multifunction $K_h(y)$ has a closed graph in $Y \times \mathcal{P}(X \times Z)$ and therefore, according to Proposition 9.1, we can find a μ-measurable map $y \mapsto (Id \times f_y^h)_{\#}\lambda_y$ such that

$$d\left((Id \times f)_{\#}\lambda_y, (Id \times f_y^h)_{\#}\lambda_y\right) = \text{dist}\left((Id \times f)_{\#}\lambda_y, K_h(y)\right) \qquad \text{for } \mu\text{-a.e. } y.$$

Defining f^h in such a way that $f^h = f_y^h$ on $\text{spt } \lambda_y$, by Step 1 we have that f^h is λ-measurable. Moreover

$$\lim_{h \to +\infty} (Id \times f^h)_{\#}\lambda_y = (Id \times f)_{\#}\lambda_y$$

for μ-a.e. y, hence (see the simple argument in the end of the proof of Theorem 7.1) we obtain that (f^h) converges in $L(X, \lambda_y; Z)$ to f for μ-a.e. y. As $\lambda = \lambda_y \otimes \mu$ we obtain that (f^h) is a Cauchy sequence in $L(X, \lambda; Z)$. Denoting by g a Borel limit function, we can find a subsequence $h(k)$ such that

$$\int_Y \sum_{k=1}^{\infty} d_{\lambda_y}\left(f^{h(k)}, g\right) d\mu = \sum_{k=1}^{\infty} d_{\lambda}\left(f^{h(k)}, g\right) < +\infty.$$

Therefore $f^{h(k)}$ converge in $L(X, \lambda_y; Z)$ to g for μ-a.e. y and $g = f$ λ_y-a.e. for μ-a.e. y.

The proof of the following measurable selection result is available for instance in [17].

Proposition 9.1 (Measurable selection). *Let $\lambda \in \mathcal{M}_+(X)$ and let $f : X \to Y$ be λ-measurable. Assume that $x \mapsto \Gamma(x)$ is a multifunction which associates to any $x \in X$ a compact and nonempty subset of Y. If the graph of Γ (i.e. $\{(x, y) : y \in \Gamma(x)\}$) is closed, there exists a λ-measurable map $g : X \to Y$ such that $g(x) \in \Gamma(x)$ and*

$$d_Y(f(x), g(x)) = \text{dist}_Y(f(x), \Gamma(x)) \quad \text{for } \lambda\text{-a.e. } x \in X.$$

In the applications to transport problems in Euclidean spaces the typical situation occurs with $B \in \mathcal{B}(\mathbf{R}^n)$ and $\pi : B \to \mathcal{S}_c(\mathbf{R}^n)$, where π satisfies with $x \in \pi(x)$ for any x. By disintegrating a measure λ concentrated on B along the level sets $\pi^{-1}(C)$ (contained in C and therefore 1-dimensional), in order to apply the 1-dimensional theory we would like to find conditions ensuring that the disintegrated measures λ_C have no atom. Although no sharp condition seems to be known, it can be shown that the absolute continuity of λ_C is inherited from λ provided the family of segments $\pi(B)$ is countably Lipschitz. The proof below is taken from [3], where this problem is discussed more in detail (see Remark 6.1 therein).

Theorem 9.4 (Absolute continuity). *Let $B \in \mathcal{B}(\mathbf{R}^n)$, let $Y = \mathcal{S}_c(\mathbf{R}^n)$ and let $\pi : B \to Y$ be a Borel map satisfying the conditions*

(i) If $\pi(x) \neq \pi(x')$ then the intersection $\pi(x) \cap \pi(x')$ can contain at most the initial point of $\pi(x)$ and of $\pi(x')$ and this point is not in B.
(ii)$x \in \pi(x)$ for any $x \in B$.
(iii)The direction $\tau(x)$ of $\pi(x)$ is a $\mathbf{S}^{n-1}$-valued countably Lipschitz map on B, i.e. there exist sets $B_h \subset B$ whose union contains B and such that $\tau|_{B_h}$ is a Lipschitz map for any h.

Then, for any measure $\lambda \in \mathcal{M}_+(\mathbf{R}^n)$ absolutely continuous with respect to $\mathcal{L}^n \llcorner B$, setting $\mu = \pi_\# \lambda \in \mathcal{M}_+(Y)$, the measures λ_C of Theorem 9.1 are absolutely continuous with respect to $\mathcal{H}^1 \llcorner C$ for μ-a.e. $C \in Y$.

Proof. Being the property stated stable under countable disjoint unions we may assume that

(a) there exists a unit vector ξ such that $\tau(x) \cdot \xi \geq \frac{1}{2}$ for any $x \in B$;
(b) $\tau(x)$ is a Lipschitz map on B;
(c) B is contained in a strip

$$\{x : a - b \leq x \cdot \xi \leq a\}$$

with $b > 0$ sufficiently small (depending only on the Lipschitz constant of ν) and $\pi(x)$ intersects the hyperplane $\{x : x \cdot \xi = a\}$.

Assuming with no loss of generality $\xi = e_n$ and $a = 0$, we write $x = (y, z)$ with $y \in \mathbf{R}^{n-1}$ and $z < 0$. Under assumption (a), the map $T : \pi(B) \to \mathbf{R}^{n-1}$ which associates to any segment $\pi(x)$ the vector $y \in \mathbf{R}^{n-1}$ such that $(y, 0) \in \pi(x)$ is well defined. Moreover, by condition (i), T is one to one. Hence, setting $f = T \circ \pi : B \to \mathbf{R}^{n-1}$,

$$\nu := T_\# \mu = f_\# \lambda, \qquad C(y) := T^{-1}(y) \supset f^{-1}(y)$$

and representing $\lambda = \eta_y \otimes \nu$ with $\eta_y = \lambda_{C(y)} \in \mathcal{M}_1(f^{-1}(y))$ (see (38)), we need only to prove that $\eta_y \ll \mathcal{H}^1 \llcorner C(y)$ for ν-a.e. y.

To this aim we examine the Jacobian, in the y variables, of the map $f(y, t)$. Writing $\tau = (\tau_y, \tau_t)$, we have

$$f(y,t) = y + d(y,t)\tau_y(y,t) \qquad \text{with} \qquad d(y,t) = -\frac{t}{\tau_t(y,t)}.$$

Since $\tau_t \geq 1/2$ and $d \leq 2b$ on B we have

$$\det\left(\nabla_y f(y,t)\right) = \det\left(Id + d\nabla_y \tau_y + \frac{t}{\tau_t^2}\nabla_y \tau_t \otimes \tau_y\right) > 0$$

if b is small enough, depending only on the Lipschitz constant of τ.
Therefore, the coarea factor

$$\mathbf{C}f := \sqrt{\sum_A \det^2 A}$$

(where the sum runs on all $(n-1) \times (n-1)$ minors A of ∇f) of f is strictly positive on B and, writing $\lambda = g\mathcal{L}^n$ with $g = 0$ out of B, Federer's coarea formula (see for instance [2], [26], [38]) gives

$$\lambda = \frac{g}{\mathbf{C}f}\mathbf{C}f\mathcal{L}^n = \frac{g}{\mathbf{C}f}\mathcal{H}^1 \llcorner f^{-1}(y) \otimes \mathcal{L}^{n-1} = \eta'_y \otimes \nu'$$

and

$$\eta'_y := \frac{\frac{g}{\mathbf{C}f}\mathcal{H}^1 \llcorner f^{-1}(y)}{\int_{f^{-1}(y)} g/\mathbf{C}f \, d\mathcal{H}^1}, \qquad \nu' := \left(\int_{f^{-1}(y)} \frac{g}{\mathbf{C}f} \, d\mathcal{H}^1\right) \mathcal{L}^{n-1}\llcorner L$$

with $L := \{y \in \mathbf{R}^{n-1} : \mathcal{H}^1(f^{-1}(y)) > 0\}$.

By Theorem 9.2 we obtain $\nu = \nu'$ and $\eta_y = \eta'_y$ for ν-a.e. y, and this concludes the proof.

Remark 9.1. As the proof clearly shows, the statement is still valid for maps $\pi : B \to \mathcal{S}_o(\mathbb{R}^n)$ provided condition (i) is replaced by the simpler condition that $\pi(x) \cap \pi(x') = \emptyset$ whenever $\pi(x) \neq \pi(x')$.

References

1. G.ALBERTI & L.AMBROSIO: *A geometric approach to monotone functions in* $\mathbb{R}^n$. Math. Z., **230** (1999), 259–316.
2. L.AMBROSIO, N.FUSCO & D.PALLARA: *Functions of Bounded Variation and Free Discontinuity Problems*. Oxford University Press, 2000.
3. L.AMBROSIO: *Lecture Notes on the Optimal Transport Problems*. Notes of a CIME Course given in Madeira (2000), to be published in the CIME Springer Lecture Notes.
4. G.ALBERTI, B.KIRCHHEIM & D.PREISS: Personal communication.
5. L.AMBROSIO, B.KIRCHHEIM & A.PRATELLI: *Existence of optimal transports with crystalline norms*. In preparation.

6. L.AMBROSIO & S.RIGOT: *Optimal mass transportation in the Heisenberg group.* In preparation.

7. G.ANZELLOTTI & S.BALDO: *Asymptotic development by Γ-convergence.* Appl. Math. Optim., **27** (1993), 105–123.

8. G.ANZELLOTTI, D.PERCIVALE & S.BALDO: *Dimension reduction in variational problems, asymptotic developments in Γ-convergence and thin structures in elasticity.* Asymptotic Anal., **9** (1994), 61–100.

9. G.BOUCHITTÉ, G.BUTTAZZO & P.SEPPECHER: *Shape optimization via Monge–Kantorovich equation.* C.R. Acad. Sci. Paris, **324-I** (1997), 1185–1191.

10. G.BOUCHITTÉ & G.BUTTAZZO: *Characterization of optimal shapes and masses through Monge–Kantorovich equation.* J. Eur. Math. Soc., **3** (2001), 139–168.

11. G.BOUCHITTÉ & G.BUTTAZZO: In preparation.

12. Y.BRENIER: *Décomposition polaire et réarrangement monotone des champs de vecteurs.* C.R. Acad. Sci. Paris, Sér I Math., **305** (1987), 805–808.

13. Y.BRENIER: *Polar factorization and monotone rearrangement of vector-valued functions.* Comm. Pure Appl. Math., **44** (1991), 375–417.

14. Y.BRENIER: *Minimal geodesics on groups of volume-preserving maps and generalized solutions of the Euler equations.* Comm. Pure Appl. Math., **52** (1999), 411–452.

15. L.CAFFARELLI: *Boundary regularity of maps with a convex potential.* Commun. Pure Appl. Math., **45** (1992), 1141–1151.

16. L.CAFFARELLI, M.FELDMAN & R.J.McCANN: *Constructing optimal maps for Monge's transport problem as a limit of strictly convex costs.* J. Amer. Math. Soc., **15** (2002), 1–26.

17. C.CASTAING & M.VALADIER: *Convex analysis and measurable multifunctions.* Lecture Notes in Mathematics **580**, Springer, 1977.

18. G.DAL MASO: *An Introduction to Γ-Convergence*, Birkhäuser, 1993.

19. C.DELLACHERIE & P.MEYER: *Probabilities and potential.* Mathematical Studies **29**, North Holland, 1978.

20. L.DE PASCALE & A.PRATELLI: *Regularity properties for Monge transport density and for solutions of some shape optimization problem.* Calc. Var., **14** (2002), 249–274.

21. L.DE PASCALE, L.C. EVANS & A.PRATELLI: In preparation.

22. H.DIETRICH: *Zur c-konvexität und c-Subdifferenzierbarkeit von Funktionalen.* Optimization, **19** (1988), 355–371.

23. L.C.EVANS: *Partial Differential Equations and Monge-Kantorovich Mass Transfer.* Current Developments in Mathematics, 1997, 65–126.

24. L.C.EVANS & W.GANGBO: *Differential Equation Methods for the Monge-Kantorovich Mass Transfer Problem.* Memoirs AMS, **653**, 1999.

25. K.J.FALCONER: *The geometry of fractal sets*, Cambridge University Press, 1985.

26. H.FEDERER: *Geometric measure theory.* Springer, 1969.

27. M.FELDMAN & R.McCANN: *Uniqueness and transport density in Monge's mass transportation problem.* 2000, to appear on Calc. Var.

28. M.FELDMAN & R.McCANN: *Monge's transport problem on a Riemannian manifold.* Trans. Amer. Mat. Soc., **354** (2002), 1667–1697.

29. W.GANGBO: *An elementary proof of the polar factorization theorem for functions.* Arch. Rat. Mech. Anal., **128** (1994), 381–399.

30. W.GANGBO & R.J.McCANN: *The geometry of optimal transportation.* Acta Math., **177** (1996), 113–161.

31. L.V.KANTOROVICH: *On the transfer of masses*. Dokl. Akad. Nauk. SSSR, **37** (1942), 227–229.

32. L.V.KANTOROVICH: *On a problem of Monge*. Uspekhi Mat. Nauk., **3** (1948), 225–226.

33. D.G.LARMAN: *A compact set of disjoint line segments in* $\mathbf{R}^3$ *whose end set has positive measure*. Mathematika, **18** (1971), 112–125.

34. R.MCCANN: *Polar factorization of maps on Riemannian manifolds*. Geom. Funct. Anal., **11** (2001), 589–608.

35. G.MONGE: *Memoire sur la Theorie des Déblais et des Remblais*. Histoire de l'Acad. des Sciences de Paris, 1781.

36. S.T.RACHEV & L.RÜSCHENDORF: *Mass transportation problems*. Vol I: Theory, Vol. II: Applications. Probability and its applications, Springer, 1998.

37. L.RÜSCHENDORF: *On c-optimal random variables*. Statistics & Probability Letters, **27** (1996), 267–270.

38. L.SIMON: *Lectures on geometric measure theory*, Proc. Centre for Math. Anal., Australian Nat. Univ., **3**, 1983.

39. C.SMITH & M.KNOTT: *On the optimal transportation of distributions*. J. Optim. Theory Appl., **52** (1987), 323–329.

40. C.SMITH & M.KNOTT: *On Hoeffding–Fréchet bounds and cyclic monotone relations*. J. Multivariate Anal., **40** (1992), 328–334.

41. V.N.SUDAKOV: *Geometric problems in the theory of infinite dimensional distributions*. Proc. Steklov Inst. Math., **141** (1979), 1–178.

42. N.S.TRUDINGER & X.J.WANG: *On the Monge mass transfer problem*. Calc. Var. PDE, **13** (2001), 19–31.

43. J.URBAS: *Mass transfer problems*, 1998.

44. C.VILLANI: *Topics in mass transportation*. Forthcoming book by AMS.

45. L.C.YOUNG: *Generalized curves and the existence of an attained absolute minimum in the calculus of variations*. Comptes Rendus Soc. Sci. et Lettres Varsovie, **30** (1937), 212–234.

46. L.C.YOUNG: *Generalized surfaces in the Calculus of Variations*. Ann. of Math., **43** (1942), 84–103.

47. L.C.YOUNG: *Generalized surfaces in the Calculus of Variations II*. Ann. of Math., **43** (1942), 530-544.

List of participants

1. Agueh Martial,
   School of Mathematics Georgia Institute of Tecnology (Atlanta), USA
   agueh@math.gatech.edu
2. Argiolas Roberto,
   Università di Cagliari , Italy
   robarg@tiscalinet.it
3. Awanou Gerard,
   University of Georgia, USA
   gawanou@math.uga.edu
4. Balaschevich Natalia,
   Institute of Mathematics National Accademy of Science Minsk, Belarus
   balash@im.bas-net.by
5. Bocea Miriam,
   Carnegie Mellon University, USA
   mbocea@andrew.cmu.edu
6. Brenier Yann (lecturer)
   CNRS, LJAD, Université de Nice, France
   brenier@math3.unice.fr
7. Buttazzo Giuseppe, (lecturer)
   Università di Pisa, Italy
   buttazzo@dm.unipi.it
8. Caffarelli Luis, (editor, lecturer)
   University of Texas, Austin, USA
   caffarel@fireant.ma.utexas.edu
9. Cerutti M.Cristina,
   Politecnico di Milano, Italy
   cricer@mate.polimi.it
10. Da Lio Francesca,
    Università di Torino, Italy
    dalio@dm.unito.it

11. De Pascale Luigi,
    Center de Mathematique Ecole Polytecnique (Palaiseau), France
    depascal@dm.unipi.it

12. Dore Giovanni,
    Università di Bologna, Italy
    dore@dm.unibo.it

13. Evans Lawrence Craig, (lecturer),
    University of Berkeley, USA
    evans@math.berkeley.edu

14. Ferrari Fausto,
    Università di Bologna, Italy
    ferrari@dm.unibo.it

15. Grimaldi Anna,
    Università di Cagliari, Italy
    grimaldi@unica.it

16. Kalligiannaki Evagelia,
    University of Crete, Greece
    evagelia@octapous.math.uch.gr

17. Loeper Gregoire,
    Université de Nice Sophia-Antipolis, France
    oeper@math.unice.fr

18. Maddalena Francesco,
    Politecnico di Bari, Italy
    maddalen@pascal.dm.uniba.it

19. Marcati Pierangelo,
    Università di L'Aquila, Italy
    marcati@univaq.it

20. Markowich Peter,
    University of Vienna (Boltzmanngasse), Austria
    Peter.Markowich@univie.ac.at

21. Maroofi Hamed,
    School of Mathematics Georgia Institute of Tecnology (Atlanta), USA
    maroofi@math.gatech.edu

22. Mikami Toshio,
    Hokkaido University (Sapporo), Japan
    mikami@math.sci.hokudai.ac.jp

23. Milakis Emmanouil,
    University of Crete, Greece
    milakis@math.uoc.gr

24. Modica Giuseppe,
    Università di Firenze, Italy
    modica@dma.unifi.it

25. Montanari Annamaria,
    Università di Bologna, Italy
    montanari@dm.unibo.it

26. Morrone Raffaele,
    Università di Torino, Italy
    raffaele.morrone@unito.it

27. Mucci Domenico,
    Università di Parma, Italy
    domenico.mucci@ipruniv.cce.unipr.it

28. Nussenzveig Lopes Helena Judith,
    UNICAMP, The State University of Campinas, Brazil
    hlopes@ime.unicamp.br

29. Paronetto Fabio,
    Università di Lecce, Italy
    fabio.paronetto@unile.it

30. Pignotti Cristina,
    Università di Roma "Tor Vergata", Italy
    pignotti@mat.uniroma2.it

31. Popovici Cristina,
    Carnegie Mellon University, USA
    cristina@andrew.cmu.edu

32. Pratelli Aldo,
    Scuola Normale Superiore Pisa, Italy
    pratelli@cibs.sns.it

33. Prinari Francesca,
    Università di Pisa, Italy
    prinari@mail.dm.unipi.it

34. Savare Giuseppe,
    Università di Pavia, Italy
    savare@ian.pv.cnr.it

35. Salsa Sandro (editor)
    Politecnico di Milano, Italy
    sansal@mate.polimi.it

36. Schneider Matthias,
    University of Mainz, Germany
    adi@mathematik.uni-mainz.de

37. Soravia Pierpaolo,
    Università di Padova, Italy
    soravia@math.unipd.it

38. Thierry Champion,
    Universite Montpelliere II, France
    champion@darboux.math.univ-montp2.fr

39. Van Goethem Nicolas,
    Université Catholique de Louvain-la-Neuve, Belgium
    vangoeth@mema.ucl.ac.be

40. Voellinger Stephan,
    Universitaet Freiburg, Germany
    voelling@stochastik.uni-freiburg.de

41. Villani Cedric (lecturer)
    UMPA, École Normale Supérieure de Lyon, France
    cvillani@umpa.ens-lyon.fr
42. Zourari Georgios,
    Ceremade Université de Paris, France
    zouraris@ceremade.dauphine.fr

LIST OF C.I.M.E. SEMINARS

1979	79. Recursion theory and computational complexity		Ed. Liguori, Napoli
	80. Mathematics of biology		&
			Birkhäuser
			"
1980	81. Wave propagation		"
	82. Harmonic analysis and group representations		
	83. Matroid theory and its applications		
1981	84. Kinetic Theories and the Boltzmann Equation	(LNM 1048)	Springer-Verlag
	85. Algebraic Threefolds	(LNM 947)	"
	86. Nonlinear Filtering and Stochastic Control	(LNM 972)	"
1982	87. Invariant Theory (LNM 996)		"
	88. Thermodynamics and Constitutive Equations (LN Physics 228)		"
	89. Fluid Dynamics	(LNM 1047)	"
1983	90. Complete Intersections	(LNM 1092)	"
	91. Bifurcation Theory and Applications	(LNM 1057)	"
	92. Numerical Methods in Fluid Dynamics	(LNM 1127)	"
1984	93. Harmonic Mappings and Minimal Immersions	(LNM 1161)	"
	94. Schrödinger Operators	(LNM 1159)	"
	95. Buildings and the Geometry of Diagrams	(LNM 1181)	"
1985	96. Probability and Analysis	(LNM 1206)	"
	97. Some Problems in Nonlinear Diffusion	(LNM 1224)	"
	98. Theory of Moduli	(LNM 1337)	"
1986	99. Inverse Problems	(LNM 1225)	"
	100. Mathematical Economics	(LNM 1330)	"
	101. Combinatorial Optimization	(LNM 1403)	"
1987	102. Relativistic Fluid Dynamics	(LNM 1385)	"
	103. Topics in Calculus of Variations	(LNM 1365)	"
1988	104. Logic and Computer Science	(LNM 1429)	"
	105. Global Geometry and Mathematical Physics	(LNM 1451)	"
1989	106. Methods of nonconvex analysis	(LNM 1446)	"
	107. Microlocal Analysis and Applications	(LNM 1495)	"
1990	108. Geometric Topology: Recent Developments	(LNM 1504)	"
	109. H∞ Control Theory	(LNM 1496)	"
	110. Mathematical Modelling of Industrial Processes	(LNM 1521)	"
1991	111. Topological Methods for Ordinary Differential Equations	(LNM 1537)	"
	112. Arithmetic Algebraic Geometry	(LNM 1553)	"
	113. Transition to Chaos in Classical and Quantum Mechanics	(LNM 1589)	"
1992	114. Dirichlet Forms	(LNM 1563)	"
	115. D-Modules, Representation Theory, and Quantum Groups	(LNM 1565)	"
	116. Nonequilibrium Problems in Many-Particle Systems	(LNM 1551)	"
1993	117. Integrable Systems and Quantum Groups	(LNM 1620)	"

Fondazione C.I.M.E.

Centro Internazionale Matematico Estivo
International Mathematical Summer Center
http://www.math.unifi.it/~cime
cime@math.unifi.it

2003 COURSES LIST

Stochastic Methods in Finance

July 6–13, Cusanus Akademie, Bressanone (Bolzano)
Joint course with European Mathematical Society

Course Directors:

Prof. Marco Frittelli (Univ. di Firenze), marco.frittelli@dmd.unifi.it
Prof. Wolfgang Runggaldier (Univ. di Padova), runggal@math.unipd.it

Hyperbolic Systems of Balance Laws

July 14–21, Cetraro (Cosenza)

Course Director:

Prof. Pierangelo Marcati (Univ. de L'Aquila), marcati@univaq.it

Symplectic 4-Manifolds and Algebraic Surfaces

September 2–10, Cetraro (Cosenza)

Course Directors:

Prof. Fabrizio Catanese (Bayreuth University)
Prof. Gang Tian (M.I.T. Boston)

Mathematical Foundation of Turbulent Viscous Flows

September 1–6, Martina Franca (Taranto)

Course Directors:

Prof. M. Cannone (Univ. de Marne-la-Vallée)
Prof.T. Miyakawa (Kobe University)

Printing and Binding: Strauss GmbH, Mörlenbach